T0372015

Guide to Environment Safety & Health Management

Industrial Innovation Series

Series Editor

Adedeji B. Badiru

Air Force Institute of Technology (AFIT) – Dayton, Ohio

PUBLISHED TITLES

Carbon Footprint Analysis: Concepts, Methods, Implementation, and Case Studies, *Matthew John Franchetti & Defne Apul*

Cellular Manufacturing: Mitigating Risk and Uncertainty, *John X. Wang*

Communication for Continuous Improvement Projects, *Tina Agustiady*

Computational Economic Analysis for Engineering and Industry, *Adedeji B. Badiru & Olufemi A. Omitaomu*

Conveyors: Applications, Selection, and Integration, *Patrick M. McGuire*

Culture and Trust in Technology-Driven Organizations, *Frances Alston*

Global Engineering: Design, Decision Making, and Communication, *Carlos Acosta, V. Jorge Leon, Charles Conrad, & Cesar O. Malave*

Guide to Environment Safety and Health Management: Developing, Implementing, and Maintaining a Continuous Improvement Program, *Frances Alston & Emily J. Millikin*

Handbook of Emergency Response: A Human Factors and Systems Engineering Approach, *Adedeji B. Badiru & LeeAnn Racz*

Handbook of Industrial Engineering Equations, Formulas, and Calculations, *Adedeji B. Badiru & Olufemi A. Omitaomu*

Handbook of Industrial and Systems Engineering, Second Edition, *Adedeji B. Badiru*

Handbook of Military Industrial Engineering, *Adedeji B. Badiru & Marlin U. Thomas*

Industrial Control Systems: Mathematical and Statistical Models and Techniques, *Adedeji B. Badiru, Oye Ibidapo-Obe, & Babatunde J. Ayeni*

Industrial Project Management: Concepts, Tools, and Techniques, *Adedeji B. Badiru, Abidemi Badiru, & Adetokunboh Badiru*

Inventory Management: Non-Classical Views, *Mohamad Y. Jaber*

Kansei Engineering—2-volume set

- Innovations of Kansei Engineering, *Mitsuo Nagamachi & Anitawati Mohd Lokman*
- Kansei/Affective Engineering, *Mitsuo Nagamachi*

Kansei Innovation: Practical Design Applications for Product and Service Development, *Mitsuo Nagamachi & Anitawati Mohd Lokman*

Knowledge Discovery from Sensor Data, *Auroop R. Ganguly, João Gama, Olufemi A. Omitaomu, Mohamed Medhat Gaber, & Ranga Raju Vatsavai*

Learning Curves: Theory, Models, and Applications, *Mohamad Y. Jaber*

Managing Projects as Investments: Earned Value to Business Value, *Stephen A. Devaux*

Modern Construction: Lean Project Delivery and Integrated Practices, *Lincoln Harding Forbes & Syed M. Ahmed*

Moving from Project Management to Project Leadership: A Practical Guide to Leading Groups, *R. Camper Bull*

Project Management: Systems, Principles, and Applications, *Adedeji B. Badiru*

Project Management for the Oil and Gas Industry: A World System Approach, *Adedeji B. Badiru & Samuel O. Osisanya*

PUBLISHED TITLES

Quality Management in Construction Projects, *Abdul Razzak Rumane*

Quality Tools for Managing Construction Projects, *Abdul Razzak Rumane*

Social Responsibility: Failure Mode Effects and Analysis, *Holly Alison Duckworth & Rosemond Ann Moore*

Statistical Techniques for Project Control, *Adedeji B. Badiru & Tina Agustiady*

STEP Project Management: Guide for Science, Technology, and Engineering Projects, *Adedeji B. Badiru*

Sustainability: Utilizing Lean Six Sigma Techniques, *Tina Agustiady & Adedeji B. Badiru*

Systems Thinking: Coping with 21st Century Problems, *John Turner Boardman & Brian J. Sauser*

Techonomics: The Theory of Industrial Evolution, *H. Lee Martin*

Total Project Control: A Practitioner's Guide to Managing Projects as Investments, Second Edition, *Stephen A. Devaux*

Triple C Model of Project Management: Communication, Cooperation, Coordination, *Adedeji B. Badiru*

FORTHCOMING TITLES

3D Printing Handbook: Product Development for the Defense Industry, *Adedeji B. Badiru & Vhance V. Valencia*

Company Success in Manufacturing Organizations: A Holistic Systems Approach, *Ana M. Ferreras & Lesia L. Crumpton-Young*

Design for Profitability: Guidelines to Cost Effectively Management the Development Process of Complex Products, *Salah Ahmed Mohamed Elmoselhy*

Essentials of Engineering Leadership and Innovation, *Pamela McCauley-Bush & Lesia L. Crumpton-Young*

Global Manufacturing Technology Transfer: Africa-USA Strategies, Adaptations, and Management, *Adedeji B. Badiru*

Handbook of Construction Management: Scope, Schedule, and Cost Control, *Abdul Razzak Rumane*

Handbook of Measurements: Benchmarks for Systems Accuracy and Precision, *Adedeji B. Badiru & LeeAnn Racz*

Introduction to Industrial Engineering, Second Edition, *Avraham Shtub & Yuval Cohen*

Manufacturing and Enterprise: An Integrated Systems Approach, *Adedeji B. Badiru, Oye Ibidapo-Obe & Babatunde J. Ayeni*

Project Management for Research: Tools and Techniques for Science and Technology, *Adedeji B. Badiru, Vhance V. Valencia & Christina Rusnock*

Project Management Simplified: A Step-by-Step Process, *Barbara Karten*

A Six Sigma Approach to Sustainability: Continual Improvement for Social Responsibility, *Holly Allison Duckworth & Andrea Hoffmeier Zimmerman*

Total Productive Maintenance: Strategies and Implementation Guide, *Tina Agustiady & Elizabeth A. Cudney*

Guide to Environment Safety & Health Management

DEVELOPING, IMPLEMENTING, & MAINTAINING A CONTINUOUS IMPROVEMENT PROGRAM

Frances Alston
Emily J. Millikin

CRC Press is an imprint of the
Taylor & Francis Group, an **informa** business

CRC Press
Taylor & Francis Group
6000 Broken Sound Parkway NW, Suite 300
Boca Raton, FL 33487-2742

Printed on acid-free paper
Version Date: 20150511

International Standard Book Number-13: 978-1-4822-5940-7 (Hardback)

Visit the Taylor & Francis Web site at
http://www.taylorandfrancis.com

and the CRC Press Web site at
http://www.crcpress.com

Contents

Preface xv
About the Authors xvii
Acronyms xix

1. Environment Safety and Health (ES&H) 1
1.1 Introduction 1
1.2 ES&H Hazards Identification and Control 1
1.2.1 Definition of the Scope of Work 2
1.2.2 Identification of Hazards 2
1.2.3 Mitigation of Hazards 4
1.3 Application of ES&H in the Design Process 5
1.4 Application in Operations and Maintenance Activities 6
1.5 Work Control Process 6
1.6 Worker Feedback 8
1.7 Organization Design 8
1.7.1 Radiation Safety 9
1.7.2 Worker Safety and Health 10
1.7.2.1 Industrial Hygiene 10
1.7.2.2 Industrial Safety 10
1.7.3 Occupational Health 11
1.7.4 Environmental Science and Protection 11
1.8 Environment Safety and Health of the Past 12
1.9 Environment Safety and Health in the Present 12
1.10 Environment Safety and Health in the Future 13
1.11 R2A2s 13
1.12 Project Management Approach to ES&H 14
1.13 The Benefits of Compliance 16
1.14 The Penalty of Noncompliance 17
1.15 The Catchall of Compliance 18
1.16 Communication 19
1.17 Summary 19

2. ES&H Organizational Structure 21
2.1 Introduction 21
2.2 Organization Structure Development 21
2.2.1 Divisional Organization Structure 22
2.2.2 Functional Organization Structure 23
2.2.3 Matrix Organization Structure 25

2.3 ES&H Organization Structure 26
2.3.1 ES&H Functional Organization......28
2.3.2 ES&H Matrix Organization......28
2.4 ES&H Organization Structure at the Department or Group Level......29
2.4.1 Radiation Protection......29
2.4.2 Environment Safety and Health Support Services......30
2.4.3 Worker Safety and Health......31
2.4.4 Environment Protection......32
2.4.5 Occupational Health Services......33
2.4.6 Environment Safety and Health Training......34
2.5 Summary......34

3. Radiation Protection 37
3.1 Introduction......37
3.2 Management and Organizational Structure......38
3.3 Roles and Responsibilities......39
3.4 Radiological Program Flow-Down of Requirements......39
3.5 Regulatory Requirements......40
3.6 Radiation Protection Programs......41
3.6.1 Radiological Control Manuals......41
3.6.2 Radiological Protection Policies and Procedures......42
3.7 Radiological Protection Program Assessment......43
3.8 Training in Radiation Protection......44
3.9 Radiation Protection Documentation......45
3.10 Complacency in the Workplace......46
3.11 Response to Abnormal Conditions......46
3.12 Balance between Being a Company and a Worker Representative...47
3.13 Summary......48

4. Worker Safety and Health 51
4.1 Introduction......51
4.2 Industrial Hygiene Programs......53
4.2.1 Asbestos......53
4.2.2 Beryllium......55
4.2.3 Biosafety......56
4.2.4 Chemical Safety and Hygiene (Hazard Communication)...56
4.2.5 Confined Space......57
4.2.6 Food Safety......57
4.2.7 Heat Stress......57
4.2.8 Lead......57
4.2.9 Nanotechnology......58
4.2.10 Hearing Conservation—Noise......58
4.2.11 LASER......58
4.2.12 Personal Protective Equipment......59

4.2.13 Refractory Ceramic Fibers 59
4.2.14 Respiratory Protection 60
4.2.15 Silica 60
4.3 Occupational and Industrial Safety 60
4.3.1 Aerial Lifts 60
4.3.2 Construction Safety 62
4.3.3 Hoisting and Rigging 62
4.3.4 Hazardous Energy 62
4.3.4.1 Lockout, Tagout 63
4.3.5 Elevators and Escalators 63
4.3.6 Ergonomics 63
4.3.7 Explosives and Blasting Agent Safety 64
4.3.8 Fall Protection 64
4.3.9 Powered Industrial Trucks 64
4.3.10 Ladders 65
4.3.11 Lighting—Illumination 65
4.3.12 Machine Guarding 65
4.3.13 Pressure Safety 66
4.3.14 Signs 66
4.3.15 Traffic Safety 66
4.3.16 Walking-Working Surfaces 66
4.3.17 Welding, Cutting, Soldering, and Brazing 67
4.4 Medical Surveillance Program 67
4.5 Occupational Exposure Limits 67
4.6 Analytical Laboratory 68
4.7 Summary 68

5. Occupational Health and Medicine 71
5.1 Introduction 71
5.2 Management and Administration 72
5.3 Functions of the Occupational Health Staff 73
5.3.1 Preemployment Physicals 74
5.3.2 Medical Surveillance 75
5.3.3 Injury and Illness Evaluation 76
5.3.4 Fitness for Duty 77
5.3.5 Return-to-Work Process 78
5.4 Prevention and Wellness Programs 79
5.5 Injury and Illness Case Management and Reporting 80
5.6 Challenges of Managing an Occupational Health Program 82
5.7 Summary 83

6. Environmental Protection 85
6.1 Introduction 85
6.2 Organization Structure and Design 85
6.3 Regulatory Structure and Drivers 88

6.4 Waste Management ... 89
6.5 Environmental Permits ... 91
6.6 Regulatory Compliance and Reporting ... 91
6.7 Environmental Sustainability ... 93
6.8 Employee Involvement ... 93
6.9 Environmental Interest Groups ... 94
6.10 Environmental Audits and Inspections ... 94
6.10.1 Environmental Management Systems ... 95
6.10.2 Compliance Auditing ... 95
6.11 Summary ... 97

7. ES&H Program Support ... 99
7.1 Introduction ... 99
7.2 Performance Assurance Office ... 100
7.3 Communication Office ... 101
7.4 Injury and Illness Management and Reporting Office ... 102
7.5 Business and Finance Office ... 103
7.6 Program Management Office ... 104
7.7 Chemical Safety ... 108
7.8 Summary ... 110

8. ES&H Training ... 111
8.1 Introduction ... 111
8.2 Systematic Approach to Training ... 113
8.2.1 Analysis of Training ... 114
8.2.2 Design of Training ... 116
8.2.3 Development of Training ... 117
8.2.4 Implementation of Training ... 118
8.2.5 Training Evaluation for Effectiveness ... 119
8.3 ES&H Training ... 120
8.4 Tracking of Training ... 120
8.5 Training Records Retention ... 122
8.6 Summary ... 122

9. Continuous Improvement of Environment Safety and Health ... 125
9.1 Introduction ... 125
9.2 Assessing the ES&H Organization ... 125
9.2.1 Assessment Techniques ... 126
9.2.2 Performance-Based Assessment Model ... 127
9.2.3 Assessment Functional Elements ... 127
9.2.3.1 Compliance Functional Element ... 128
9.2.3.2 Effectiveness Functional Element ... 128
9.2.3.3 Quality Functional Element ... 129
9.2.4 Performance-Based Assessment Plan ... 130
9.2.4.1 Assessment Plan Objectives ... 130
9.2.4.2 Schedule of Performance ... 131

9.2.4.3 Identification of Assessor(s) 131
9.2.4.4 Assessment Criteria and Lines of Inquiry 132
9.2.5 Assessment Reporting and Results 132
9.2.5.1 General Assessment Information 132
9.2.5.2 Signature Approval Page 133
9.2.5.3 Summary of Assessment Techniques and Observations 133
9.3 Performance Metrics and Performance Indicators 134
9.3.1 Performance Metrics 134
9.3.2 Performance Indicators 135
9.3.3 Leading Metrics and Indicators 138
9.3.4 Lagging Metrics and Indicators 138
9.3.5 Qualities of Solid Performance Metrics and Indicators 139
9.3.6 Performance Metrics and Indicators in the Continuous Improvement Process 140
9.3.7 Use of ES&H Metrics and Performance Indicators in Company Management 143
9.4 Summary 144

10. Project Management Approach to Environment Safety and Health 145
10.1 Introduction 145
10.2 The Project Management Approach 145
10.3 ES&H Scheduling 146
10.4 Managing Cost 152
10.5 ES&H Managers as Project Managers 152
10.5.1 Leadership Skill 154
10.5.2 Team Building 154
10.5.3 Differing Opinion Resolution 155
10.5.4 Technical Knowledge 156
10.5.5 Resource Allocation and Management 156
10.5.6 Organizational Skills 156
10.5.7 Planning Skills 156
10.6 Time Management 157
10.7 Summary 159

11. Succession Planning 161
11.1 Introduction 161
11.2 Employee Retention Strategy 162
11.3 The Role of Management in the Succession Planning Process 163
11.4 Attributes of a Good Succession Plan 164
11.4.1 Identify Key Positions 165
11.4.2 Identification of Key Competencies 166
11.4.3 Identification of Candidates 167
11.4.4 Candidate Communication 168

11.4.5 Gap Analysis ... 169
11.4.6 Define the Training and Development Plan ... 170
11.4.7 Implementation of the Training and Development Plan ... 173
11.4.8 Evaluate Succession Plan ... 173
11.5 The Role of a Mentor in Succession Planning ... 174
11.6 External Hiring Process ... 175
11.7 ES&H Organizational Succession Strategy ... 175
11.8 Summary ... 177

12. Technology and the ES&H Profession ... 179
12.1 Introduction ... 179
12.2 Use of Technology in the Workplace ... 180
12.2.1 Search Engines ... 180
12.2.2 Regulations and Standards ... 180
12.2.3 Applications Used by the ES&H Discipline ... 181
12.2.4 Communications and Graphics ... 182
12.2.5 Emergency Response ... 182
12.2.6 Employee Observational Programs ... 183
12.2.7 Training and Certification ... 184
12.3 Types of Technological Devices ... 184
12.3.1 Smartphones ... 184
12.3.2 Laptops and Tablets ... 185
12.3.3 Portable Instrumentation ... 185
12.3.4 Simulation Devices and Digital Imaging ... 186
12.4 Technology Considerations ... 187
12.4.1 Organization and Storage of Media ... 187
12.4.2 Validation of Data Used for Reporting ... 187
12.4.3 Security Protection of Electronic Data ... 188
12.4.4 Use of Electronic Devices in the Workplace ... 188
12.4.5 Social Media Sites and Networking within a Technological Work Environment ... 188
12.4.6 Planning Costs Associated with Technological Devices ... 189
12.4.7 Overcoming Generational Differences ... 189
12.5 Cost–Benefit Analysis ... 190
12.6 Summary ... 192

13. Culture in the ES&H Work Environment ... 193
13.1 Introduction ... 193
13.2 Impact of Culture in the ES&H Environment ... 194
13.2.1 Safety and Health ... 194
13.2.2 Radiological Protection ... 195
13.2.3 Environmental Protection ... 196

13.3 Methods to Evaluate Culture 196
13.3.1 One-on-One Daily Discussions 197
13.3.2 Interviews 198
13.3.2.1 Individual Interviews 198
13.3.2.2 Group Interviews 198
13.3.2.3 General Considerations When Interviewing ... 199
13.3.3 Surveys 199
13.3.4 Performance Indicators 200
13.4 Culture Improvement Initiatives within ES&H 201
13.4.1 Defined Culture Values 201
13.4.2 Key Performance Indicators 201
13.4.3 Leadership and Supervisor Training 202
13.4.4 Communications Plan 204
13.4.5 Focused Campaigns 204
13.4.6 Employee Involvement Committees 205
13.5 Tools for Improving Culture 205
13.5.1 Organizational Culture Questionnaire 206
13.5.2 Culture Improvement Plan 207
13.6 Summary 207

14. The Impact of Trust in an ES&H Organization 209
14.1 Introduction 209
14.2 Trust the Building Block for Organizational Success 210
14.3 The Impact of Mistrust in an Organization 210
14.4 The Role of Trust in an ES&H Organization 211
14.5 How to Establish Trust with Customers 213
14.5.1 Humility 215
14.5.2 Relationship 215
14.5.3 Openness and Honesty 215
14.5.4 Concern for Employees 215
14.5.5 Competence 215
14.5.6 Identification 216
14.5.7 Reliability 216
14.6 The Role of the ES&H Leadership Team in Building Trust 216
14.7 The Corporate Safety Culture and Trust 217
14.8 Assessing Organizational Trust 219
14.9 Summary 220

References 223

Index 225

Preface

The field of environment safety and health (ES&H) pertains to the management and control of hazards to personnel, the environment, and the community. Developing, implementing, and managing an effective ES&H program is complex due to the vast amount of regulatory and contractual requirements that may be imposed on businesses. Regulatory applicability depends on the type of business, product produced, and potential impacts to employees, the public, and the environment. Additionally, the perception exists with some business owners and executives that the rules and regulations imposed or enforced do not directly add to the financial bottom line. A solid ES&H program can ensure regulatory compliance and contribute to the success of the company both monetarily and by shaping public perception of the company. An ES&H program that is strategic and follows project management concepts can add to the bottom line in many ways; however, the exact financial gain oftentimes cannot be quantified in the near term and in hard dollars.

When establishing and managing an ES&H organization, there are key programs and topical areas that must be addressed in order for the organization, and ultimately the company, to be successful:

- Organizational structure and succession planning
- Fundamental understanding of ES&H functional areas
- Training
- The approach and measurement of continuous organizational improvement
- Project management of ES&H
- Application of technology
- Culture and trust in the workplace

This book covers the primary areas of ES&H and key elements that should be considered in developing, managing, and implementing the program. It is intended to serve as a practical guide for ES&H managers and professionals to use in executing a successful program.

About the Authors

Frances Alston, PhD, has built a solid career foundation over the past 25 years by leading the development and management of environment safety, health, and quality (ESH&Q) programs in diverse cultural environments. Throughout her career, she has delivered superior performances within complex, multistakeholder situations and has effectively dealt with challenging safety, operational, programmatic, regulatory, and environmental issues.

Dr. Alston has been effective in facilitating integration of ESH&Q programs and policies as a core business function while leading a staff of business, scientific, and technical professionals. She is skilled in providing technical expertise in regulatory and compliance arenas as well as determining necessary and sufficient program requirements to ensure employee and public safety, including environmental stewardship and sustainability. She also has extensive knowledge and experience in assessing programs and cultures to determine areas for improvement and development of strategy for improvement.

Dr. Alston earned a BS in industrial hygiene and safety, an MS in hazardous and water materials management/environmental engineering, an MSE in systems engineering/engineering management, and a PhD in industrial and systems engineering. She is a fellow of the American Society for Engineering Management (ASEM) and holds certifications as a Certified Hazardous Materials Manager (CHMM) and a Professional Engineering Manager (PEM). Dr. Alston's research interests include investigating and implementing ways to design work cultures that facilitate trust.

Emily J. Millikin has more than 29 years of leadership experience in regulatory, environmental compliance, radiation protection, and safety and health programs at multiple Department of Energy (DOE) and Department of Defense (DOD) sites. She has served as safety, health, and quality director and subject matter expert for ESH&Q programs. She has extensive leadership experience in safety, health, and quality programs and has managed all aspects of program, cost, and field implementation of safety and health, industrial hygiene, radiological control, environmental, quality assurance, contractor assurance system, emergency preparedness, safeguards and security, occupational health, and Price–Anderson Amendment Act programs. Millikin has achieved over 11 million safe work hours while consistently demonstrating a low total recordable case rate. She has established successful employee-led behavioral-based safety observation programs and successfully achieved Voluntary Protection Program (VPP) Star status.

Millikin earned a BS in environmental health from Purdue University with double majors in industrial hygiene and health physics. She is a Certified Safety Professional (CSP), Safety Trained Supervisor (STS), and certified in the National Registry for Radiation Protection Technologists.

Acronyms

ACA	Affordable Care Act
ACGIH	American Conference of Government Industrial Hygienists
ACM	Asbestos-contaminated material
ADA	Americans with Disabilities Act
AEC	Atomic Energy Commission
ALARA	As low as reasonably achievable
ANSI	American National Standards Institute
CAA	Clean Air Act
CD	Compact disc
CDC	Centers for Disease Control and Prevention
CERCLA	Comprehensive Environmental Response, Compensation, and Liability Act
CFR	Code of Federal Regulations
CHP	Certified Health Physicist
CIH	Certified Industrial Hygienist
CSP	Certified Safety Professional
CTIM	Culture trust integration model
CWA	Clean Water Act
D&D	Decontamination and decommissioning
DOE	U.S. Department of Energy
EFA	Environmental functional area
EMS	Environmental management system
EP	Environmental protection
EPA	U.S. Environmental Protection Agency
EPCRA	Emergency Planning and Community Right-to-Know Act
ERDA	Energy Research and Development Administration
ES&H	Environment safety and health
FWS	Fish and Wildlife Services
HI&C	Hazards identification and control
HIPAA	Health Insurance Portability and Accountability Act
ID	Task identification
IH	Industrial hygienist
ISO	International Organization for Standardization
IT	Information technology
JHA	Job hazard analysis
KPI	Key performance indicator

LASER	Light amplification stimulated emission of radiation
LOI	Line of inquiry
LSO	Laser safety officer
MIC	Methyl isocyanate
MSDS	Material safety data sheet
NFPA	National Fire Protection Association
NHPA	National Historic Preservation Act
NIOSH	National Institute of Occupational Safety and Health
NMFS	National Marine Fisheries Services
NRC	U.S. Nuclear Regulatory Commission
NRRPT	National Registry of Radiation Protection Technologists
OCQ	Organizational culture questionnaire
OEL	Occupational exposure limit
OSH	Occupational safety and health
OSHA	Occupational Safety and Health Administration
PEL	Permissible exposure limit
PI	Performance indicator
PM	Project management
PMIIM	Performance metric and indicator improvement model
PPE	Personal protective equipment
QA	Quality assurance
R2A2	Roles, responsibilities, accountability, and authority
R&D	Research and development
RCF	Refractory ceramic fiber
RCO	Radiological control officer
RCRA	Resource Conservation Recovery Act
RI	Responsible individual
SARA	Superfund Amendments and Reauthorization Act
SAT	Systematic approach to training
SDS	Safety data sheet
SME	Subject matter expert
SOMD	Site occupational medical director
STEM	Science, technology, engineering, and math
TRC	Total recordable case
TSCA	Toxic Substances Control Act
TWA	Time-weighted average
USACE	U.S. Army Corps of Engineers
WSH	Worker safety and health

1

Environment Safety and Health (ES&H)

1.1 Introduction

The environment safety and health profession is a broad and diverse field primarily governed by regulations that can result in monetary penalties if not followed. The rubrics or disciplines included in ES&H are designed primarily to protect the health of workers and the environment from adverse effects resulting from the workplace. A well-designed and implemented program can serve as a means to also protect the company from frivolous claims and accusations that may result from disgruntled employees. In addition, an effective ES&H organization is designed to support the identification and control of workplace hazards. The success of an ES&H program is predicated upon the ability to attract and sustain the staff with the appropriate skills and knowledge. Therefore, it is important to have a succession strategy that is complementary to the goal of the organization and resource needs.

Typical areas of ES&H will include radiation safety, industrial hygiene, conventional or industrial safety, occupational health, and environmental science and protection. Other areas that can be included in an ES&H organizational structure, which depends upon the goal of the organization, are fire protection, nuclear safety, emergency management, security, and quality assurance. A detailed discussion of each of these areas can be found in the chapter written specifically for that discipline to provide detailed information on program design and implementation.

1.2 ES&H Hazards Identification and Control

Fundamental to an ES&H organization is the hazards identification and control (HI&C) process. The basic premise of an HI&C process is that a logical and consistent approach exists whereby industrial safety, industrial hygiene, radiological, and environmental hazards are identified and mitigated. The generic hazards and control process consists of basically three elements:

- Definition of the scope of work
- Identification of hazards
- Mitigation of hazards

Figure 1.1 depicts a generic hazards and control process.

1.2.1 Definition of the Scope of Work

The definition of the scope of work is the first element of the hazards and control process and forms the foundation by which hazards are identified and mitigated. Work may be defined through either a requirement or a need. For example, the Occupational Safety and Health Administration (OSHA) requires a worker to perform a thorough periodic inspection of alloy steel chain slings in use on a regular basis, or if a piece of equipment has broken down and is in need of repair. In both examples, the scope of the work must be identified and defined.

Once the work is bound, then steps must be defined to determine how to complete the work. The method for defining the steps needed to complete a task may be formal or informal; however, without the discrete work steps being defined, it is not possible to determine who will be required to accomplish the work, how the work will be performed, or materials needed to perform the job.

The last step in defining the scope of work is to identify the resources that will be needed to complete the work steps. Depending upon the complexity of the work scope, several types of personnel may be required, such as an electrician, operator, carpenter, or plumber, or the work may be performed just by one person who has the skills and knowledge to ensure the work is done completely and safely.

1.2.2 Identification of Hazards

The identification of hazards should be performed for each discrete work step that was defined and should address all hazards:

- Industrial safety (i.e., falling surfaces, fire hazard)
- Chemical (i.e., organic, inorganic)
- Biological (i.e., bacterial, virus, human and animal pathogens)
- Radiological (radiation exposure and contamination)
- Environmental (i.e., toxins and pollutants, waste, and natural disasters)

When performing the hazards identification process, the regulations should be consulted because there may be specific requirements associated with a particular hazard, such as the process safety requirements or waste

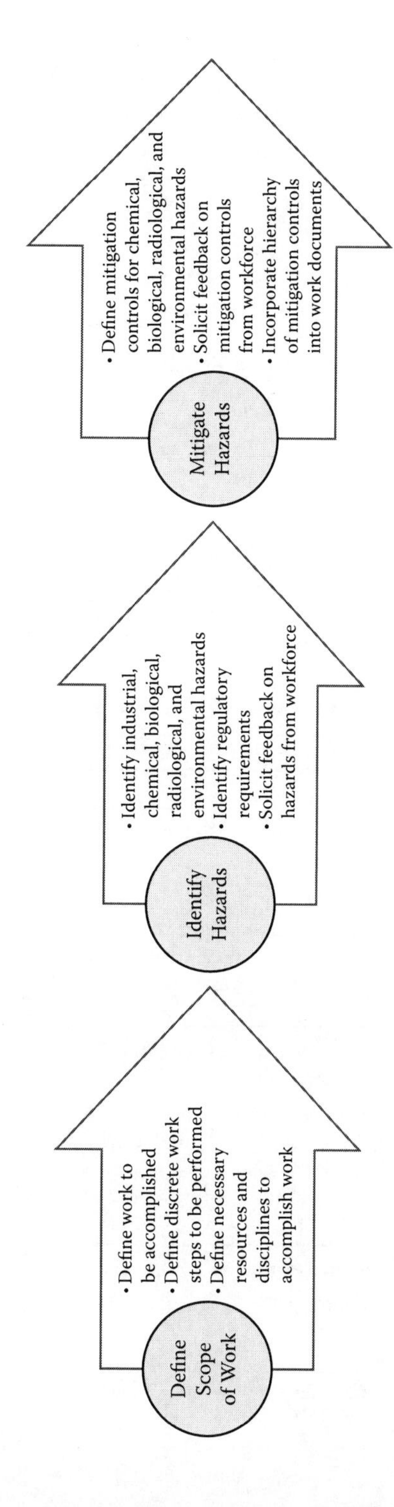

FIGURE 1.1
Hazards and control process.

disposal requirements. It is also important to enlist the assistance from workers who have or will be performing the work because they are most familiar with the hazards to be encountered, and for the most part are most familiar with the environmental surroundings.

1.2.3 Mitigation of Hazards

When defining mitigating controls for hazards there is a specific hierarchy that should be followed:

- Hazard elimination/product substitution
- Engineering controls
- Administrative controls
- Personal protective equipment (PPE)

Most of the federal regulations require use of the hierarchy of controls shown in Figure 1.2.

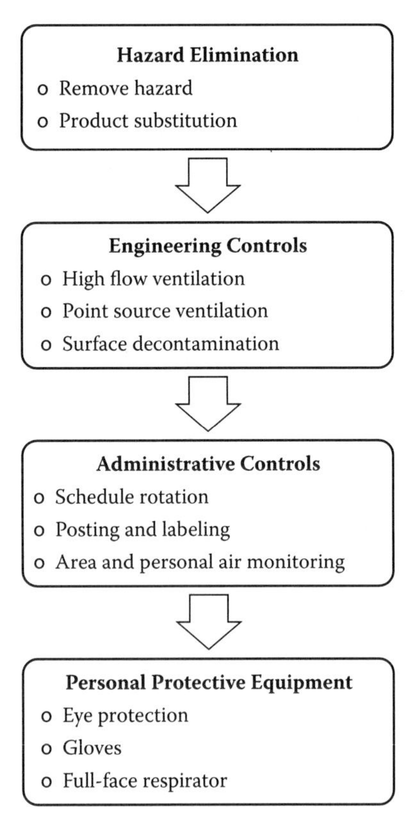

FIGURE 1.2
Hierarchy of hazard controls.

The primary reason for performing the hazards identification and mitigation process is to protect the worker and environment; however, these steps are also regulatory driven. The regulatory drivers for the ES&H function require the identification and mitigation of hazards in the workplace. According to OSHA, Section 5(a)(1), "an employer has a legal obligation to provide a workplace free of conditions or activities that either the employer or industry recognizes as hazards and that cause, or are likely to cause death or serious physical harm to employees when there is a feasible method to abate the hazard." This requirement is more commonly known as the general duty clause and is the basis by which specific industrial hazards are mitigated. According to the Nuclear Regulatory Commission (NRC), 10 CFR 20.1101(b), "The licensee shall use, to the extent practical, procedures and engineering controls based upon sound radiation protection principles to achieve occupational doses and doses to members of the public that are as low as is reasonably achievable (ALARA)."

The Environmental Protection Agency (EPA) has many parts to the overall regulatory structure; however, 40 CFR Chapter 1, Subchapter A, Part 1, Subpart A, states, "The U.S. Environmental Protection Agency permits coordinated governmental action to assure the protection of the environment by abating and controlling pollution on a systematic basis. Reorganization Plan 3 of 1970 transferred to EPA a variety of research, monitoring, standard and enforcement activities related to pollution abatement and control to provide for the treatment of the environment as a single interrelated system."

1.3 Application of ES&H in the Design Process

There is a very distinct role for the ES&H professional or manager in the design of process or operational systems. Many of the regulations require that the process of hazard identification and mitigation be applied as part of equipment or system design. This is most relevant during development of a new operational system or new facility. The role of the ES&H professional or manager is to be actively engaged as part of the design team identifying and mitigating hazards while a product or facility is in its conceptual stage. If proficient, the ES&H professional can many times eliminate the hazard by making the engineers aware of how their design can impact the worker and the environment.

The ES&H professional or manager should not assume that the engineers designing a system or facility have ever worked in the field. In many cases, the engineers are skilled at design (and many vary by discipline, such as electrical, mechanical, chemical, etc.), but may not be aware of how work is actually conducted. Therefore, feedback that the ES&H professional or manager provides can prove to be invaluable during conceptual design.

It is also recommended that the ES&H professional or manager document recommendations made during the design process. Documented feedback can prove to be valuable in the future when regulatory, operational, or maintenance personnel question how hazards were mitigated as part of the design process. Often these recommendations are useful when undergoing readiness reviews and operating the systems. As is the case in the radiological protection function, these reviews are documented as ALARA design reviews; however, any format that is used for documenting hazard mitigation strategies is acceptable.

1.4 Application in Operations and Maintenance Activities

Most often the ES&H professional or manager is directly involved in the planning and performance of operations and maintenance activities. Operational activities may include installation and running of equipment, routine tasks, or operating an equipment panel. The definition of operations will vary depending upon the service or equipment the company provides. Maintenance activities are generally divided into two categories: preventive maintenance and corrective maintenance.

Preventive maintenance is associated with any form of routine inspection or servicing that is conducted to ensure equipment is operating at its optimum level. An example of a preventive maintenance activity is changing the oil on your personal car. Corrective maintenance are actions performed to fix or repair equipment that has failed. An example of a corrective maintenance activity is replacement of your brakes on your personal car. Both preventive and corrective maintenance activities are required to maintain operations.

The ES&H manager and professional are intimately involved in executing operations and maintenance activities on a daily basis. Whether scheduling resources to support the daily activities, supporting work in the field, monitoring and responding to conditions in the field, or participating in postjob reviews, the role of the ES&H manager and professional is considered an integral part of the operations and maintenance team.

1.5 Work Control Process

The work control process is the system by which work is organized for execution. It is considered the heart of how work is conducted. Through the work control system, scope of work is defined, planned, scheduled, executed, and

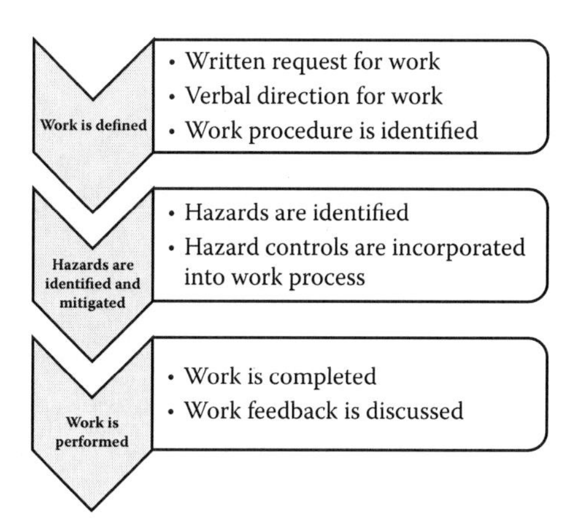

FIGURE 1.3
Work control process.

closed out. Figure 1.3 depicts a simplistic approach to a work control process. There are primarily three elements of a simplistic work control process.

Work is defined. The scope of the work to be performed and the manner by which the work is defined will vary depending upon the operations or maintenance to be performed; the work may simply consist of a supervisor directing an employee to go move a box, or it can be as complex as a work package containing multiple tasks. The manner by which work is defined depends upon the number of steps that will need to be performed associated with that particular scope of work and the level of risk to the employee and the company.

Hazards are identified and mitigated. Following the definition of work, hazards associated with performing and completing the work are identified and mitigated. The identification and mitigation of hazards may be as simple as talking to the worker and discussing what to watch out for when performing the work, or it could be as formal as a job hazards analysis (JHA) that is completed with the work crew. The identification and mitigation of hazards should follow the hierarchy of elimination, engineering, administrative, or personal protective equipment (PPE). Not all of these methods for mitigating hazards shown in Figure 1.3 are used, but the preferred method to mitigate hazards is to either eliminate or engineer the hazards out of the work process.

Work is performed. Once the work has been defined and hazards identified and mitigated, work is ready to be performed. The performance of work may simply be a supervisor directing a worker to go move a forklift while ensuring that the worker is trained and qualified to perform the task, or it can be as complex as performing a formal prejob, ensuring all workers understand their roles and responsibilities, and are formally directed to go out into the workplace and complete the defined tasks of a complex work package. If the scope of work is significant, the performance of work may take several days, weeks, or months to complete.

1.6 Worker Feedback

Some of the most useful information to be gained by the ES&H manager or professional is to listen to feedback from the worker. Worker feedback is important because

- The worker is most familiar with how the work is conducted.
- The worker generally has a better understanding of hazards that may be present in association with the task, and he or she oftentimes has ideas on how to be mitigate them.
- They have a good understanding of how to best communicate hazards and mitigation controls with the work crew.

Worker feedback is one of the primary means by which an ES&H manager and professional can successfully support completion of work. More often than not the worker has previously performed the work and understands what methods for completion of the work are successful and those methods that will not work. The worker is the "first line of defense" in understanding what hazards are present and the best method to mitigate the hazards. Depending upon the work crew and their previous level of work experience, the level of hazard mitigation required will vary and may be significant or simply make the worker(s) aware of the hazard.

One of the biggest mistakes the ES&H manager and professional can make is to meet with the workers after a job has been completed and communicate his or her perception of how the work was conducted. The worker and work crew are most familiar with the work environment and to communicate to them that you know more can be misconstrued as being disrespectful. Getting the workers involved in the work planning and mitigation process will prove to be useful to the work planner, the worker, and ultimately the company.

1.7 Organization Design

An important attribute to ensure success of an ES&H organization is to confirm that the organization design meets the needs of the company or project. Organization structures are necessary to enable the performance of work. An organization structure can facilitate or hamper the flow of business and the quality of support provided to the customer.

The reporting structure for the ES&H organization tends to be most effective when the ES&H manager reports directly to the president of the company.

Reporting to the office of the president allows a level of independence for the ES&H manager and professional that helps avoid the perception that environmental safety and health practices and policies are not in support of and in the best interest of the company, the workers, and the environment. It also removes the perception of the lack of independence in decision making and program and policy implementation.

1.7.1 Radiation Safety

The area of radiation safety for discussion in this book includes ionizing and nonionizing radiation (with the exception of lasers). Lasers will be covered in Chapter 4. Most workers have a great respect for the potential hazards associated with ionizing radiation. Oftentimes the radiological workers are more willing to follow the policies, regulations, and safe work practices that are established to keep them safe and prevent adverse health effects. Because of the high respect for the associated hazards and the training that is required to qualify personnel for their jobs, workers are not willing to take chances. However, on the other hand, workers may not be as vigilant when working with sources of nonionizing radiation. An effective radiation safety program should include the following at a minimum:

- Management commitment for program development and implementation
- Highly skilled professionals to support and implement every aspect of the program
- Succession strategy to maintain qualified professionals
- A regulatory compliant program that is documented and revised as changes occur
- Training for workers with the appropriate level and frequency of retraining
- Radiation monitoring equipment
- An equipment calibration program (can be internal or external)
- Process to evaluate program effectiveness and make changes as necessary
- Personnel dose monitoring and reporting
- Record keeping

It is incumbent upon management to develop and implement a radiation safety program that will protect workers from sources of radiation hazards in the workplace. Management is also responsible for ensuring workers receive training on how to work safely with radiation-generating sources that are present in their work environment.

1.7.2 Worker Safety and Health

More and more today the areas of industrial hygiene and conventional safety are combined into one area called worker safety and health (WSH). This makes sense because the two areas often overlap in the hazard recognition and control process. In addition, having a professional trained to perform both functions is a more efficient means of delivering services. When designing an ES&H organization it is more economical from a cost and efficiency perspective to combine the areas of industrial or conventional safety and industrial hygiene. Therefore, it is pertinent to seek professionals with knowledge in both areas or develop an onboarding training strategy to develop the necessary skills so that the professional can provide support to both areas.

1.7.2.1 Industrial Hygiene

The field of industrial hygiene is commonly defined as the field that is responsible for anticipating, recognizing, and controlling environmental stressors and workplace hazards. The industrial hygiene professional focuses on some specific physical, chemical, and biological type hazards. The field of industrial hygiene uses scientific and engineering methodologies to determine potential hazards and evaluate exposure risks and implement controls. These hazards can include chemicals, ergonomics, ionizing or nonionizing radiation, heat stress, blood-borne pathogens, or noise.

Industrial hygienists are typically well trained and are skilled in the practice of hazard recognition, mitigation, and control. In many cases these professionals hold certification as a Certified Industrial Hygienist (CIH). The role of an industrial hygienist can include the following:

- Developing sampling strategies for exposure to various chemicals
- Identifying and recommending control measures for safety-related issues
- Developing health and safety programs and assisting in implementation
- Recommending corrective measures to eliminate workplace hazards
- Participating in the recommendation and design of engineering control
- Recommending personal protective equipment
- Evaluating and interpreting regulatory requirements
- Interpreting, analyzing, and communicating exposure data

1.7.2.2 Industrial Safety

The area of industrial safety focuses on safety of the worker and the environment. As mentioned earlier, the area of industrial hygiene and industrial safety can overlap in some areas depending on the structure of the programs and the organization. Industrial safety professionals are also skilled in the recognition, evaluation, and control of workplace hazards. However, their primary focus is on the physical hazards that may occur. Like industrial hygiene professionals, many industrial safety professionals are skilled and knowledgeable in safety program development and implementation and hold certifications as a Certified Safety Professional (CSP). The role of a safety professional can include the following:

- Identifying and recommending control measures for safety-related issues
- Developing safety programs and assisting in implementation
- Recommending corrective measures to eliminate workplace hazards
- Participating in the recommendation and design of engineering control
- Recommending personal protective equipment
- Evaluating and interpreting regulatory requirements

1.7.3 Occupational Health

Providing for occupational health services is an integral part of a proficient health and safety program. Many of the OSHA standards require medical surveillance for workers to ensure that the work being performed and the chemicals that are present do not have an adverse impact on the health of the worker. The type and level of medical staff needed to support a viable occupational health program can be determined by several factors, including the following:

- The type of hazards present in the workplace
- The type of services provided (internal and external)
- The number of expected employees to service

Many companies use their health services capabilities to assist workers in maintaining a healthy lifestyle through offering programs to improve nutrition, exercise, and other health-related activities. The benefits of a workplace wellness program may include increased productivity, improved morale among workers, decreased injury and illness rates, and reduced health care costs.

1.7.4 Environmental Science and Protection

The field of environmental science and protection is also driven primarily by regulations enacted by federal, local, and state governmental agencies to ensure protection of the environment. These regulations are necessary since past activities by many companies have resulted in negatively impacting the environment. The typical ES&H professional possesses degrees in the areas of science, engineering, statistics, and the medical field. These highly skilled professionals are crucial in interpreting environmental regulatory requirements, collecting data to demonstrate compliance, and assisting in performing work within the regulations that are applicable to the business.

More detail on developing and implementing a good environmental protection program will be discussed in Chapter 6. A well-structured environmental protection program can pay for itself many times over through the avoidance of fees associated with fines resulting from noncompliance.

1.8 Environment Safety and Health of the Past

The field of environment safety and health has greatly evolved over the past century. With the development of the Occupational Safety and Health Administration (OSHA) in 1971 workplace safety has greatly improved. OSHA promulgated standards to cover all workplaces to ensure that workers have a safe place to work. A marked decrease in fatalities and injuries has been witnessed over the past decade. Safety and health for the workplace grew significantly as an area of importance during the 1970s with the passing of the Occupational Safety and Health Act (OSH Act) of 1970. After the passing of the OSH Act, companies began developing ES&H organizations and programs to aid in compliance with the act. As a result, workplaces are much safer and fewer injuries and deaths are occurring.

1.9 Environment Safety and Health in the Present

ES&H organizations are an integral part of an effective corporate program. Typically, ES&H professionals are integrated in every operational aspect of the business. This provides an opportunity for the ES&H professional to have an impact on every stage of a project or task. Early involvement of an ES&H professional can have huge benefits for the company and the health and safety of the workforce, as well as increase efficiency in task performance. Present ES&H organizations tend to be lean, having minimal resources to develop and implement programs as well as provide technical support. Therefore, it is important to develop lean programs that are based on the needs of the company and the applicable regulatory requirements.

1.10 Environment Safety and Health in the Future

Care and consideration must be given to the goals and objectives of the project along with the financial aspects of delivering the needed support when designing an ES&H organization to meet future needs of a project and company. Many organizations have found themselves in a position of trying to accomplish more with less due to budget constraints. This trend is expected to continue since funding in some industries is tighter than in previous years. Oftentimes when budgets are diminishing, support functions such as ES&H are the first to be reduced. Therefore, it is important for the needs of the company to take on a balanced approach when it comes to the professional skills and resource needs and having a viable and functioning ES&H program that is able to attract and retain the skilled workers needed.

1.11 R2A2s

R2A2 is a term commonly used to refer to roles, responsibility, accountability, and authority. It is important that R2A2s be established, documented, and communicated. When the R2A2s are clear and understood, it is easier for people to perform in the expected manner, collaboration is enhanced, and buy-in for the company goals, objectives, and business strategy is facilitated.

ES&H implementation in an organization can be a little contentious if organizational members do not know where the responsibility for ES&H resides. Many would say that the ES&H staff has responsibility for ES&H. This statement is true to a degree since the ES&H team has responsibility for knowing what regulations apply to the business, including developing policies and procedures for implementation for those applicable regulations and standards. On the other hand, implementation of the ES&H program is the responsibility of line management and the employees. The global point is that the environment safety and health of the company and its employees are the responsibility of the entire team (management and employees). Therefore, the success of an ES&H program resides with the entire company, organization, or project team. Primary responsibility resides with senior management since they control every aspect of the business, to include resources and funding.

Employers are responsible for providing a safe workplace for employees. Managers are hired to represent the employer, therefore assuming the responsibility for safety of the workforce. OSHA mandates that employers are responsible for the following:

- They must provide a workplace that does not have known serious hazards.
- They must follow the OSHA safety and health standards as they applies to the workplace.
- They must identify and correct all workplace hazards.
- They must try to eliminate or reduce hazards by first making changes in conditions as opposed to dependence on personal protective equipment.
- They must substitute hazardous chemicals for less hazardous, where feasible.

Accountability and transparency for the health and safety of every worker are a prime responsibility of the management team. Not only is it important to establish accountability, but also it must be seen by others in the way the organization conducts business daily.

1.12 Project Management Approach to ES&H

A productive ES&H program should be developed and implemented using a project management life cycle approach. The approach involves having ES&H considerations at all levels of the project planning, initiation, execution, and closure process, as shown in Figure 1.4.

Taking a project management approach is a practical way to ensure that an ES&H program is designed to meets the needs of the project or process that it is designed to support. Project management tools such as scheduling and tracking of commitments, tasks, or small project initiatives can make the difference between ensuring compliance or losing track of important regulatory requirements or initiatives to facilitate compliance. The project management model can be used in the following manner as an example:

Planning stage. The planning stage of a project or task is the optimal time to involve the ES&H professional to begin analysis of the potential impact of the proposed work tasks on the worker, stakeholders, and environment. During this stage, potential hazard recognition and mitigation can be included in the project design, and if environmental permits are necessary, they can be identified and the application process can be initiated. In addition, many regulations require considerations of ES&H principles as a part of the design process. These up-front activities can save time and money, ensure compliance with regulations, and improve the safety of the project and the subsequent tasks that will be performed by workers.

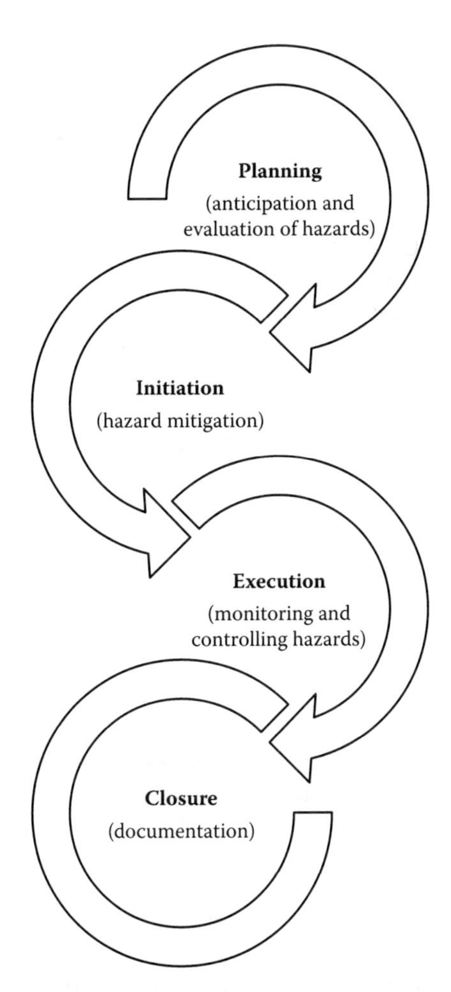

FIGURE 1.4
ES&H project management life cycle approach.

Initiation. Once the project has been developed and testing is being initiated this provides yet another opportunity for the ES&H team to review and make determination of potential hazard mitigation that may be present as a result of changes made during the initiation phase.

Execution. Many or most of the hazards should have been addressed during the planning and initiation phase. However, there are times when hazards that are identified during the two stages cannot be completely addressed. There are times when the most practical time to address a hazard is during the execution phase. The ES&H professional has yet another opportunity to analyze the remaining potential hazards and recommend controls and, in some cases, the appropriate personal protective equipment to protect workers.

Closure. The ES&H team is also involved with the project during the closing phase. When the project is complete and there is minimal physical work being performed, there may be limited involvement, such as ensuring the documentation is complete and stored in a place that it can be retrieved in the future if necessary. In the closure phase of a project this is the time where deactivation and demolition (D&D) activities take place. During D&D activities there can be a host of unique hazards present that were not present during the other three phases of the project. Therefore, a complete hazard analysis is warranted to protect the workers and the environment from adverse impact.

The project management approach to ES&H will be discussed in detail in Chapter 10.

Proper ES&H integration in the planning phase can eliminate issues related to cost, schedule, and management of the project during the operational phase.

1.13 The Benefits of Compliance

Regulatory compliance refers to an organization's adherence to and compliance with the laws and regulations that are relevant to its business. The laws can be local, state, or federal, and cover a variety of compliance issues. Violations of local, state, or federal regulations can often result in a variety of legal punishments, including fines or time in prison for egregious disregard of a regulation. A company that demonstrates a history of being regulatory compliant can serve as a means to give customers the confidence they need in a company's ability to follow through and deliver contractually and ethically when conducting business.

Management. Management can benefit from operating in a compliant environment by gaining the trust and support of the employees, customers, stakeholders, regulators, and surrounding community. When a management team and a company have a reputation for operating based on regulatory requirements, the regulators may choose not to inspect as often based on the company's compliance track record.

Employee. In a compliant work environment, employees are more confident that the workplace is safe and provides them the opportunity to conduct their work in a manner that allows them to return home safely at the end of the day. A compliant and safe work environment can serve as a vehicle in increasing productivity, morale, and trust in the leadership team and the organization.

The environment. There are many regulations designed to protect the environment from negative impacts resulting from the conduct of business. In a compliant work environment the risk of irreversible damage to the environment significantly decreases, along with fines imposed by regulatory governing agencies. Responsible environmental stewardship is key in gaining and retaining the trust and support from the employee, customer, and stakeholders.

The stakeholders. When stakeholders believe that the company has embraced the responsibility for ensuring the safety and health of the workforce and the environment, they are more likely to be at ease with the business operations and are more willing to provide support when needed. For example, when a new or revised environmental permit is needed, during the public hearing stage the risk of objections from the community as a whole can be minimized if the company has historically demonstrated responsibility and concern for the health and well-being of the workforce and the community.

1.14 The Penalty of Noncompliance

A company that operates in a noncompliance posture can inflict negative impacts on the workers, the environment, the stakeholders, and the management team. The penalty for noncompliance to an ES&H regulation may include the following:

- Large monetary fines
- Damage to company reputation
- Decreased ability to obtain projects in their specified business line
- Increased regulatory oversight

These penalties can cost the company greatly, especially in many ways, since all penalties, although not monetary, will impact the financial bottom line for the company.

Management. In an environment where compliance to ES&H regulations is not valued or adhered to, management tends to lose the trust of the workforce. Workers value and support managers and supervisors that demonstrate that they have the well-being of the worker in mind. When management does not demonstrate adherence to regulatory requirements, employees may commit themselves to performing the bare minimum and are not willing to go the extra mile when needed. In addition, it becomes difficult for the worker to invest and commit to the company long term; therefore, employee turnover may become excessive, and in some cases the management team may be involved in an increased amount of legal litigation.

Employee. In an environment where a company is consistently operating in a state of noncompliance, employees tend to believe that management and the company are not concerned about their well-being, which leads to a lack of trust. Eventually, employees may begin making their own decisions on how work is conducted to ensure they are legally compliant.

The environment. The environment can be heavily impacted from noncompliance with environmental standards. The release of many chemicals and products to the environment can produce a level of contamination to the environment that can render the soil contaminated and unusable, the air unsafe for humans to breathe, and kill plants and animals. Noncompliance to environmental standards can result in large monetary penalties and damage to the reputation of the company and the leadership team.

The stakeholders. Penalties imposed for regulatory noncompliance can degrade the financial bottom line and stock prices for the company. Depending on the severity of the noncompliance, a large penalty can be assessed that will take away from the profit margin for the company and impact the amount paid to stakeholders. Noncompliance can also deplete trust in the company and its management team in having the ability to operate a safe and compliant operation that does not adversely impact the workers and the environment.

1.15 The Catchall of Compliance

The primary objective of the OSHA regulations is to protect workers from suffering adverse health effects resulting from conducting work in the workplace. The OSHA regulations cover a large majority of the potential workplace hazards in various parts of the regulation documented in 29 CFR 1910 and 29 CFR 1926. However, don't assume that OSHA is not applicable in cases where there is a hazard and a standard is not written to deal specifically with the workplace hazard. The general duty clause places responsibility on the employer to provide a workplace that is protective of the worker. Subsequently, employers are obligated to institute whatever means needed to abate hazards when they are present. If an employer fails to comply, OSHA can issue a citation under the general duty clause.

1.16 Communication

Not only is it necessary for the ES&H management team and professional to be skilled in the technical aspects of their jobs, but they must also be skilled and compassionate communicators. Oftentimes the communication of the information, if not delivered in the appropriate manner, can lead to unrest, distrust, and produce anxiety to the worker and the community. When communicating with workers, customers, and stakeholders it is important that the following be considered:

- The type of message that is being delivered and the potential impact
- The potential emotional impact of the message
- The best method to use in delivering the message (for example, written or verbal)

Development of good communication skills is not generally focused on during college years when studying science, technology, engineering, and math (STEM); therefore, it is incumbent upon the employer to develop the skills needed to deliver messages of various types to the workforce. Delivering the message to an employee that a potential overexposure occurred during the performance of a task requires both technical knowledge and good communication skills if the message is to be believed. Workers must trust that the information provided by the ES&H staff is accurate. Therefore, ES&H management and professionals must have a sufficient level of communication skills to deliver various types of messages and skillfully perform their jobs.

1.17 Summary

A well-designed and implemented ES&H program can have lasting impacts on a company. Alternatively, a poorly designed and implemented program can also have lasting negative effects on a company. The benefits of having an effective ES&H program include the following:

- A healthier workforce
- Potential lower insurance rates

- Reduction in injury and illness rates
- Less inspection visits from regulators
- Reduced risk of noncompliances
- Fewer fees assessed by regulatory agencies for noncompliances
- Support from the workforce, stakeholders, and community

ES&H involvement in planning work integrates the practical application of hazard identification and control into the work planning and execution process. Just as with any corporation, an effective program begins with an appropriate policy statement and the support of the leadership team.

2

ES&H Organizational Structure

2.1 Introduction

The structure of an organization is an important element of a business that provides formality and consistency in program and task implementation that is necessary to enable a company to successfully achieve its mission. An organizational structure is highly dependent upon the objective and the strategy of the company. The structure can help determine the culture, communication channels, and how effective and efficient an organization can be in conducting business and providing service to its customers. It also determines the decision-making mechanisms used and the roles, responsibility, and authority of the leadership team.

Recognizing that no one organizational structure is perfect, there are several types of structures that can be used to meet the needs of an organization. These types of structures include divisional, functional, and matrix. Organizations having discrete business units are more suitable for operating in a divisional type structure, whereas the functional organization structure is aligned based on discrete activities. In a matrix structure the organization is divided based on functions across several divisional lines. A matrix organizational structure is generally used in and more effective in large organizations. Each of these structures has its own set of benefits and pitfalls. Later in this chapter we will cover the structures commonly used in designing an ES&H organization.

2.2 Organization Structure Development

The structure of an organization is a key element in providing formality and consistency in implementation of a company's business strategy. Therefore, it is important that great care and consideration be given during structure design. An organizational structure should be developed taking into consideration the following:

- The goal of the company and organization.
- The strategy used in business interactions.
- The scope to be performed.
- The size of the organization.

There are many ways an organization can be structured to accomplish the same mission and strategy; all structures will present some limitations or complexities during implementation. The goal in designing a proper organizational structure is to minimize the complexities and limitations and its impact on the performance of work. The types of structures that are most often used during the design of a typical ES&H organization are discussed in the subsections below and shown in Figures 2.1 to 2.4. The organizations that are represented by these structures will be discussed in detail later in this chapter.

2.2.1 Divisional Organization Structure

An organization structured along divisional lines is designed with the purpose of grouping products or functions into divisions. Each division is focused on only supporting one product or project at a time and can help facilitate smooth operation and focus on the project at hand. There are some distinct advantages as well as disadvantages to a divisional organization structure. Advantages include the following:

- Little or no impact will been seen by other divisions in the event of failure of one division.
- Better coordination of support.
- Knowledge sharing is facilitated.
- The ability to focus on a single project or product.
- Clear lines of communication between jobs within the department.
- The chain of command is clear and easy to follow.

Disadvantages include the following:

- Inefficiencies in duplication of efforts resulting from separating specialized functions.
- Staff performing the same functions may not communicate with others in other divisions, which can create inconsistencies in implementation of like programs within the company.
- Difficult to develop, facilitate, and sustain one corporate culture.
- Difficult for the department to understand and support the goal of the company due to operating in isolation.

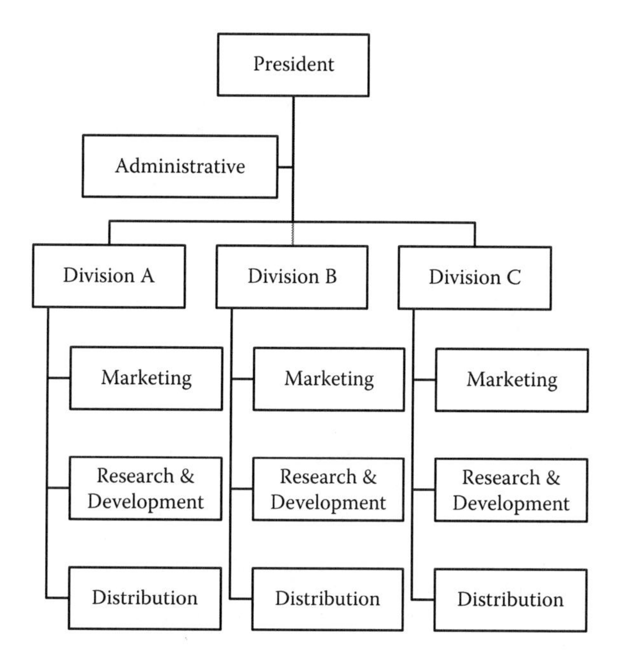

FIGURE 2.1
Divisional organization structure.

- The management team can become myopic and put the goals of the department before the company goals.
- Increase cost to a company for support type services.

A typical divisional organizational structure is shown in Figure 2.1. When an ES&H program is structured along divisional lines, one can expect to have a larger staff to provide the focus support, as opposed to sharing of resources that would occur in other types of organizational structures. A pure divisional structure does not seem to be as prevalent in many companies today because sharing of resources is discouraged. Sharing of resources is a popular protocol used to reduce cost and increase consistency across a company.

2.2.2 Functional Organization Structure

In a functional organization products and projects are organized based on functions performed. In this structure decision making is decentralized and delegated to the functional unit responsible for the project or product. Departments and functions that are typically best suited for a functional organization structure include the following:

- Accounting and finance department
- Engineering department

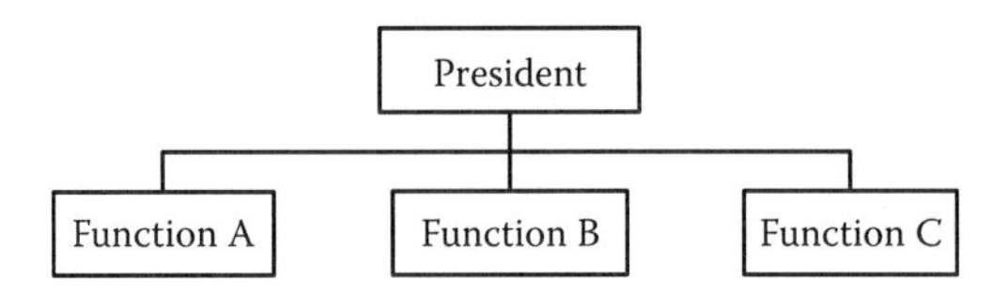

FIGURE 2.2
Functional organization structure.

- Human resources department
- Legal department
- Marketing department
- Facilities and infrastructure department
- Public relations department
- Production department
- ES&H programmatic functions
- Purchasing department

A typical functional structured organization is shown in Figure 2.2.

The functional organizational structure shares some of the same advantages and disadvantages of a divisional structure. Advantages and disadvantages of a functional structure are listed below. Advantages include the following:

- Staff is most often managed by a person with experience in their same specialty that is able to better evaluate work and make informed decisions.
- Knowledge sharing is facilitated.
- Increased efficiencies in operations and process flow because the employees are allowed to focus on one specific functional area.
- Clear lines of communication between jobs within the department.
- The chain of command is clear and easy to follow.
- Gives employees a sense of belonging.
- Easier to develop and monitor career paths for employees.

Disadvantages include:

- Difficult for the department to understand and support the goal of the company due to operating in isolation.
- The management team can become myopic and put the goals of the department before the company goals.
- Potential limited opportunities for advancement since staffers are limited to functional area activities, which makes competing with colleagues for advancement more difficult.

- Difficulties working and communicating with other functional areas since they tend to perform tasks and conduct business within their function.
- The size of the functional areas can become difficult to manage because they are seemingly isolated and they can begin functioning like a small internal company with their own culture and unique processes.
- Functional areas tend to focus on their own goals rather than on overall company objectives.
- The hierarchical nature of the functional structure creates bureaucracy that can impede change.

2.2.3 Matrix Organization Structure

In a matrix structure the organization is divided based on functions across several organizational lines and individuals are grouped by function or product line. The matrix organization structure is used in large companies. The decision-making process in a matrix structure is typically used in large companies.

In a centralized matrix structure decisions tend to be made by a few people that are at the top of the organization. On the other hand, in a decentralized structure decision-making authority is dispersed throughout the company. A typical matrix structure is shown in Figure 2.3. Advantages and disadvantages of a matrix structure are listed below.

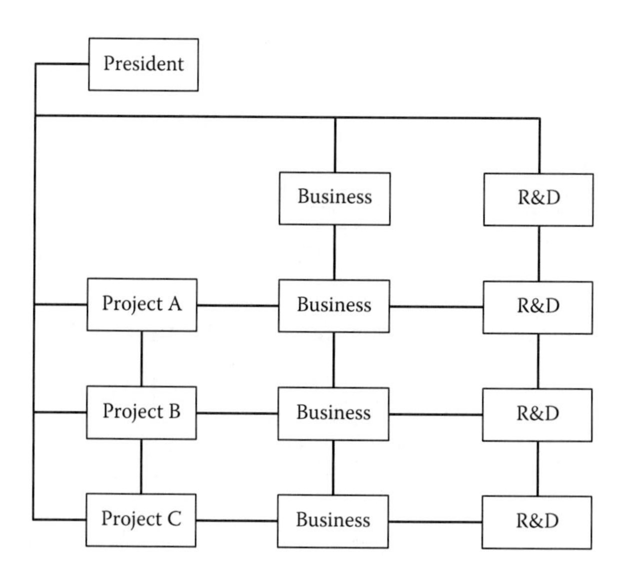

FIGURE 2.3
Matrix organization structure.

Advantages include the following:

- More efficient use of resources since resources can be shared across projects.
- Increase in the flow of information both across and up through the organization.
- Employees are in contact with many people across several departments, which helps provide career opportunities with other departments within the company.
- Potential for cross-training of personnel among the disciplines and job positions.

Disadvantages include the following:

- Increased complexity in the chain of command, adding to the worker–manager ratio.
- Matrix structure employees may be faced with conflicted loyalties.
- Can be expensive to maintain due to the need to have duplication of management leading essentially the same functions.

The matrix structure is most commonly used for an ES&H organization in large companies to deliver support to the various divisions and product lines. In smaller companies a functional structure is often used due to scope and reduced staffing level.

2.3 ES&H Organization Structure

The type of work being performed by the organization, along with applicability of the various regulatory requirements, dictates the makeup of the environment safety and health (ES&H) organization. These elements can also be used to assist in "right sizing" the organization during design. A typical functional structure for an ES&H organization is shown in Figure 2.4.

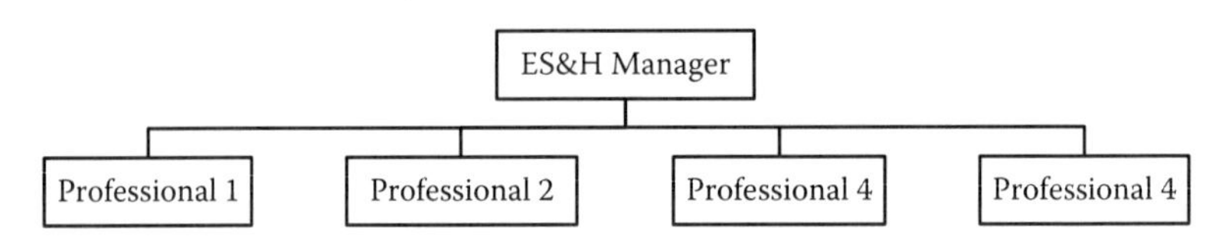

FIGURE 2.4
ES&H organization structure—small project.

ES&H support is frequently delivered by the functional organization where the scope and objective of the company does not lend itself to the need of a large ES&H structure. The following characteristics may justify the structure of a functional organization:

- The organization based on its scope and objectives does not trigger the applicability of many of the regulatory requirements.
- The company is small and centralized.
- The company has a small staffing level.

When designing an ES&H organization for a small company or project a simple functional model may make sense from a distribution and funding standpoint. In such case, one would want to hire professionals that are skilled in more than one aspect of ES&H. Multiskilled professionals increase efficiency in getting work accomplished and reduce the amount of staffing needed to perform the functions of ES&H. For example, when reviewing a project for ES&H applicability, one or two professionals that have a diverse knowledge base may be enough to support project needs instead of three or four professionals with specialized experience and knowledge. All of the applicable regulatory requirements are still required to be implemented even in small companies. Therefore, the structure must be designed to ensure that all applicable regulatory requirements are addressed. The organizational design may resemble the structure in Figure 2.4.

In the case where the organization or company is large, an ES&H structure can be designed a little differently to account for the size of the organization as well as taking into account the scope and regulatory driver, as shown in Figure 2.5. In this structure all of the workers report to a manager in the functional organization and provide support from the functional organization or through a matrix relationship to the project or other organization within the company. This also means that the day-to-day direction will be provided by the home organization management team. Also, limited direction may be provided through a matrix manager.

FIGURE 2.5
ES&H organization structure—large project.

2.3.1 ES&H Functional Organization

The backbone of an ES&H program development and implementation is the functional portion of the organization. The functional portion of ES&H serves as the vehicle to

- Develop policies and practices to ensure implementation of ES&H regulatory requirements
- Communicate with the regulatory agencies on behalf of the company
- Collect, analyze, and report data to demonstrate compliance
- Monitor and advise the workforce on avenues to facilitate employee health
- Keep management and workers apprised of regulatory changes and impact to the company

2.3.2 ES&H Matrix Organization

One of the most common organization structures for an ES&H organization is a matrix structure. A matrix structure can provide the best of both worlds for ES&H professionals and the groups that they support. This structure allows the professional to report to the organization or group in which he or she provides day-to-day support. Yet he or she is connected to the home or functional ES&H organization, as shown in Figure 2.6. This connection to the home organization is essential in ensuring consistency of service

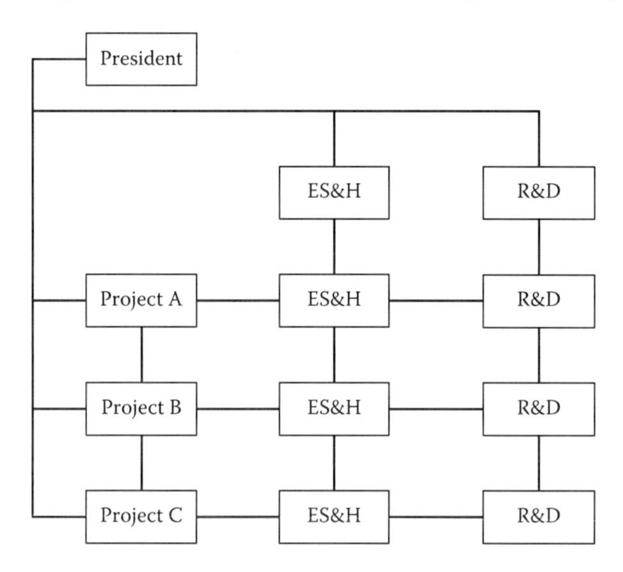

FIGURE 2.6
Matrix organization structure.

delivery, program implementation, and independence in everyday decision making in the field.

2.4 ES&H Organization Structure at the Department or Group Level

As indicated in the organization structure for the ES&H organization at the highest level, the organization is a multifaceted department with varying functions. Each group shares one important element in common, which is to enable safe performance of work. The structure of each group is just as important as the structure of the entire department in order to ensure formality and efficiency in meeting the goals of the group and the department. Each of these groups has distinct functions, yet they complement each other in ways to provide complete and total support for the health and safety of the worker and the environment.

2.4.1 Radiation Protection

The radiation protection group is found only in companies where radioactive equipment or materials are handled. The typical radiation protection program for most companies is small because handling large amounts of radioactive products or equipment containing or producing radiation is not the norm for most industries. However, companies that are competing in the nuclear industry have large encompassing radiation safety programs and groups. These groups tend to have state-of-the-art measuring and monitoring equipment along with a highly skilled technical staff. The organization structure in Figure 2.7 represents a program designed for an industry that typically handles moderate to large amounts of radioactive materials and sources. It is not uncommon to have one individual responsible for radiation safety in a small organization or an organization with very little hazards from radiation sources and materials.

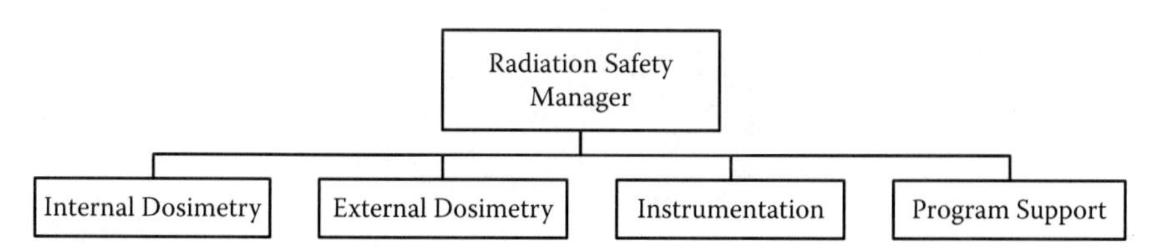

FIGURE 2.7
Radiation protection.

Industries that would typically have a need for a radiation protection program include the following:

- Nuclear industry
- Medical (hospitals)
- Instrumentation calibration where a radiation source is necessary to perform calibration
- Radiography
- Radioactive sources
- Research and development
- Academic and scientific

2.4.2 Environment Safety and Health Support Services

ES&H support services groups most frequently comprise all of the support type functions that are necessary to support the core functions of the ES&H organization. It is most effective if these functions are a part of a group to enhance efficiency in operations and provide opportunities for cross-training and job sharing of key support functions. The typical structure for this group is found in Figure 2.8.

The support services group of ES&H is most proficient when included as a part of the functional organization structure of the global ES&H organization. Moving these functions into a matrix structure can deplete consistency and efficiencies in operations and reporting. The support services group provides a wide array of services, to include the following:

- Injury and illness reporting
- Assessments of programs and implementation
- Procurement
- Procedure writing

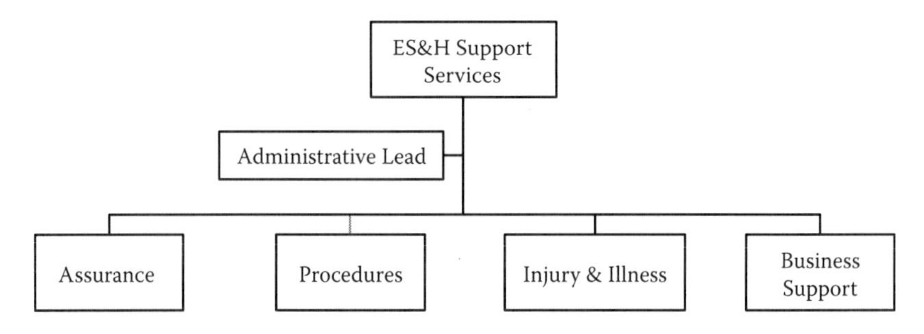

FIGURE 2.8
ES&H support services.

- Safety and health communication
- Budget tracking and analysis
- Metrics development and analysis
- Document control

The assurance portion of the group is responsible for monitoring the ES&H organization performance against the content of its contracts with the customers and regulatory requirements. Monitoring occurs in the form of conducting assessments and providing feedback to the organizations on implementation performance. Metrics are also used to gauge and track performance at the group, department, and company levels.

2.4.3 Worker Safety and Health

The worker safety and health (WSH) group is the area of ES&H where safety policies and practices must be embraced and understood by all in order to be effective. It is by far the most common area of ES&H and the one that most people are familiar with because the discipline pertains to the management of physical, chemical, and biological hazards. A typical organization structure for the WSH group is shown in Figure 2.9.

The field of industrial hygiene is primarily governed or managed by the Occupational Safety and Health Administration (OSHA). Additional standards that are applicable to the field of industrial hygiene include those written by the National Fire Protection Association (NFPA) and the American National Standards Institute (ANSI). These standards dictate that a program be in place that protects workers from various hazards generated in the workplace. As a result, it is necessary that subject matter experts be employed and trained to serve as experts in providing consultation on standard interpretation and implementation. Many of the areas that must be considered are listed in Table 2.1.

The industrial safety program is similar to the industrial hygiene program in that it too is governed by regulatory requirements such as OSHA, NFPA, and ANSI. In order to ensure program implementation, it is important to structure implementation around assigning various subject matter experts

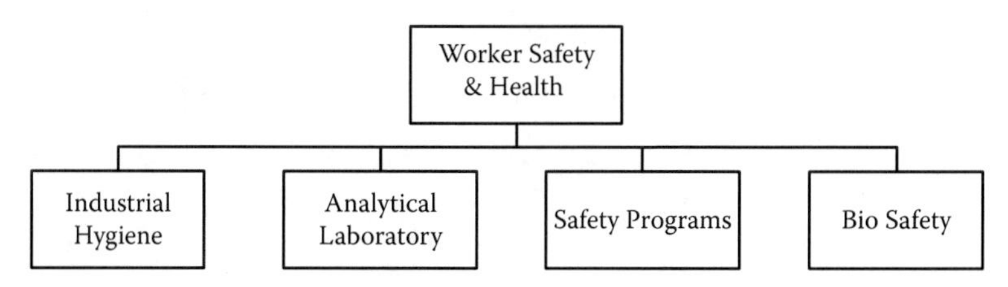

FIGURE 2.9
Worker safety and health.

TABLE 2.1
Example Industrial Hygiene Programs

Asbestos	Food safety
Beryllium	Hazard communication
Biosafety	Heat stress
Chemical hygiene	Lead
Chemical safety	Nanotechnology
Confined space	Noise
Nonionizing radiation	Personal protective equipment—industrial hygiene
Silica	Respiratory protection
Ventilation	Refractory ceramic fibers

TABLE 2.2
Example Industrial Safety Programs

Aerial lifts	Fall protection
Aviation safety	Firearms
Construction safety	Fork trucks
Cranes/hoisting/rigging	Ladders
Electrical safety	Lasers
Elevators	Lighting
Ergonomics	Lockout, tagout
Explosives	Machine guarding
Personal protective equipment—industrial safety	Pressure safety (including compressed gases, cryogens)
Signs	Traffic safety
Walking surfaces	Welding

who are required to maintain knowledge in various aspects of the regulatory requirements. It is common for one individual to serve as SME for several programs or regulatory requirements. However, on the other hand, it is not practical for one individual to maintain adequate knowledge in all areas to adequately support the entire company, especially if the company is diverse in functions and staff. Many program areas are outlined in Table 2.2.

2.4.4 Environment Protection

The environmental protection group is essential in ensuring compliance to a wide array of regulatory requirements. A simple organization chart is shown in Figure 2.10. These requirements can have grave negative impacts to the environment if not appropriately implemented. Regulations are written

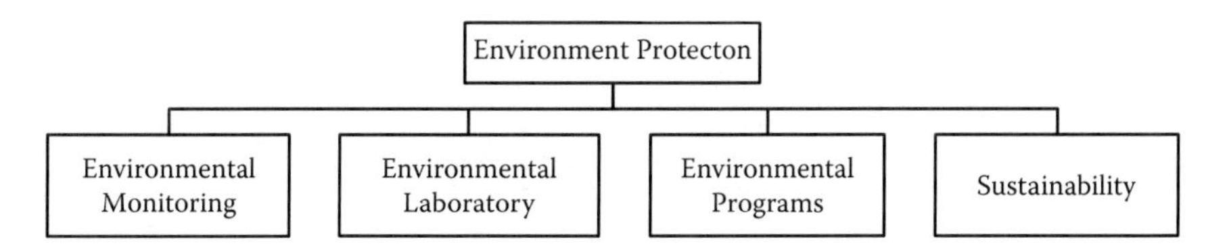

FIGURE 2.10
Environmental protection structure.

for environmental areas that are prescriptive in compliance requirement. These areas include the following:

- Pollution prevention
- Cultural resources
- Air
- Water
- Waste
- Environmental sustainability

2.4.5 Occupational Health Services

The primary purpose of the occupational health services staff is to ensure the health and well-being of the workers. They are also responsible for determining if a worker is healthy enough to perform assigned tasks. The primary services generally provided by the health services staff include the following:

- Determining fitness for duty
- Determining if employee is drug- and alcohol-free
- Determining if an employee is capable of performing tasks associated with his or her job
- Determining any restrictions for employees performing certain tasks
- Conducting medical monitoring for medical surveillance program
- Determining psychological stability for work performance

The leadership model can be a shared model between the site occupational medicine doctor (SOMD) and the health services clinic manager. The SOMD is primarily responsible to ensure the delivery of the appropriate medical care. The health services manager is responsible for all leadership aspects to ensure appropriate infrastructure and staffing are available to ensure clinic operation and delivery of services. An example of an organization chart for the health services group is shown in Figure 2.11.

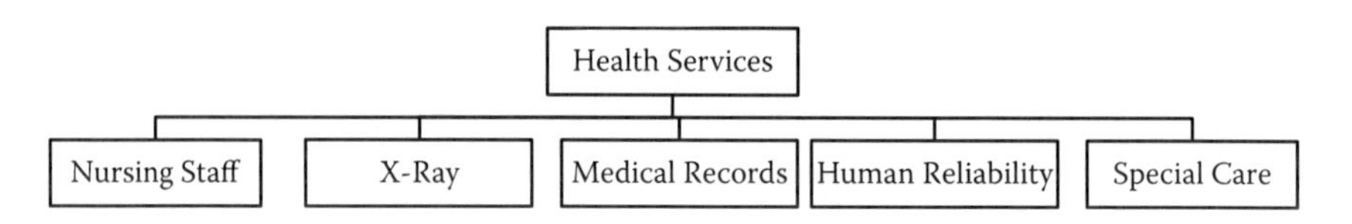

FIGURE 2.11
Health services.

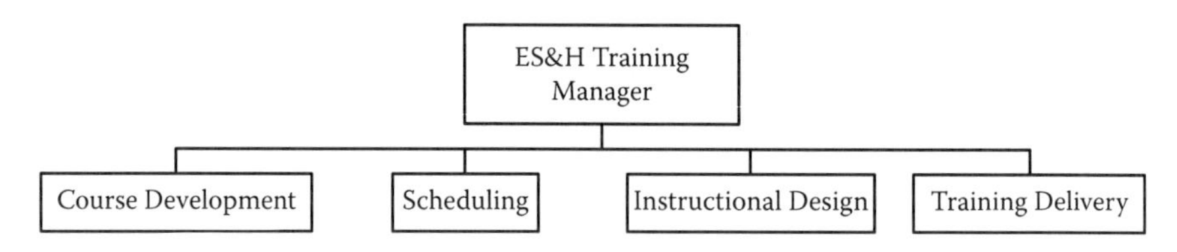

FIGURE 2.12
ES&H training organization structure.

2.4.6 Environment Safety and Health Training

Training is an important aspect of any company strategy to obtain and maintain knowledge and skills required to perform the business of the company. Training is also an integral part of implementing a succession planning strategy. Most of the required continuing education type of training performed in most companies is associated with ES&H. As such, it is important to have a functioning group with the capability to design and deliver courses to maintain knowledge and compliance with the many regulatory requirements. A typical training group structure is shown in Figure 2.12.

2.5 Summary

The structure of an organization is important to ensure consistency in implementing business practices and interacting with the customer. The organization structure can also play a part in shaping the culture and developing trust among its members. When designing the structure for the overall organization, considerations must be given to the goal, mission, and strategy of the company or project.

The structure of an ES&H organization is important to the success of program implementation and to be in a position to serve the needs of its customer. Keeping in mind that essentially any organization structure can work as long as those that have a vested interest want it to succeed, it is difficult to gain efficiencies in implementation of ES&H services if the organization design is to provide "stovepipe" support to projects and groups along with increased risk for regulatory noncompliance. The typical ES&H organization is aligned with both a functional organization and a matrix structure. The functional aspect of the organization is commonly designed to provide programmatic support to enable work to be performed safely and in a compliant manner. The functional portion of the organization typically is most effective in focusing on program development and management elements such as

- Procedure development
- Continuous improvement and feedback (assessments, metrics)
- Policy development
- Regulatory interpretation and compliance

The matrix portion of the organization is primarily focused on program implementation. This segment of the organization staff is generally matrixed to and integrated into the programs or project that they provide technical support to. The matrixed ES&H staff is essential in ensuring and enabling regulatory compliance and implementation of the policies and procedures developed by the functional organization to facilitate safety and health of the workforce, the community, and the environment.

3

Radiation Protection

3.1 Introduction

The field of radiation protection is the science and practice of protecting people and the environment from the harmful effects of ionizing radiation. Ionizing radiation is widely used in industry and medicine and can present a significant health hazard if exposure is not controlled and managed. Ionizing radiation, through either external or internal exposure, can cause damage to living tissue and bones, skin burns, radiation sickness, and death at high exposure levels. Exposure to low levels of radiation is also known to statistically elevate the risk of a person developing cancer. Over the years, the study of radiation protection has evolved.

After the Manhattan Project began producing fissionable material in 1942, many new sources and levels of radiation hazards were observed and the focus of radiation protection began to emerge. Prior to this time, scientists were primarily concerned with protecting themselves and their coworkers. With testing associated with the atomic bomb for national defense, these concerns were extended to protecting the general public and the environment (Health Physics Society). As concern grew, so too did the need to regulate the industry and form a consistent approach to monitoring and protecting people and the environment.

The U.S. Atomic Energy Commission (AEC) was an agency of the U.S. government established after World War II by Congress to foster and control the peacetime development of atomic science and technology. President Truman signed the McMahon/Atomic Energy Act on August 1, 1946, transferring the control of atomic energy from the military to civilian hands, effective from January 1, 1947. Transferring the control of atomic energy to a nonmilitary body—control of the production plants, laboratories, equipment, and personnel assembled during the war to produce the atomic bomb—started the formation of how the commercial nuclear industry would exist today.

The AEC played a key role in establishing regulations and providing financial resources for promoting the peaceful use of atomic energy. In addition, the AEC provided crucial financial resources to study the effects of radiation on the environment. By the 1960s an increasing number of critics charged that the AEC's regulations were insufficiently rigorous in several important areas, including the following:

- Radiation protection standards
- Nuclear reactor safety
- Plant siting
- Environmental protection

By 1974, the AEC's regulatory programs were criticized and Congress proceeded to abolish the AEC with its functions assigned to two new agencies: the Energy Research and Development Administration (ERDA) and the Nuclear Regulatory Commission (NRC). In 1977 President Carter signed into law the Department of Energy Organization Act of 1977, which created the Department of Energy (DOE). Today the NRC and DOE are managed as two distinct government agencies.

The NRC oversees nuclear business and research, oversight of nuclear medicine, and nuclear safety. The DOE manages the development and oversight of nuclear weapons, and provides research and promotion of civil uses of radioactive materials and nuclear power. In addition, other agencies self-regulate radiation protection, such as the Army, Navy, and Air Force, to name a few. However, this chapter will focus on the two largest agencies: the NRC and the DOE.

3.2 Management and Organizational Structure

The radiation protection organization is generally organized into two areas: program and field execution. Organizationally, program personnel support policy and procedure development, act as subject matter experts (SMEs), and prepare technical bases documents necessary to support field execution. In addition, there is an oversight function, either as part of the radiological protection organization or separate, that performs audits and assessments to ensure compliance with regulatory requirements. An example of a typical radiation protection organization is shown in Figure 3.1.

Personnel who act as radiological protection professionals generally have health physics degrees and some form of certification, such as a Certified Health Physicist (CHP) or National Registry of Radiation Protection Technologists (NRRPT). If the radiological protection manager does not directly hold this type of pedigree, then personnel who directly support the manager should have the technical education and certifications. In both the NRC and DOE environments, that level of pedigree is expected and required.

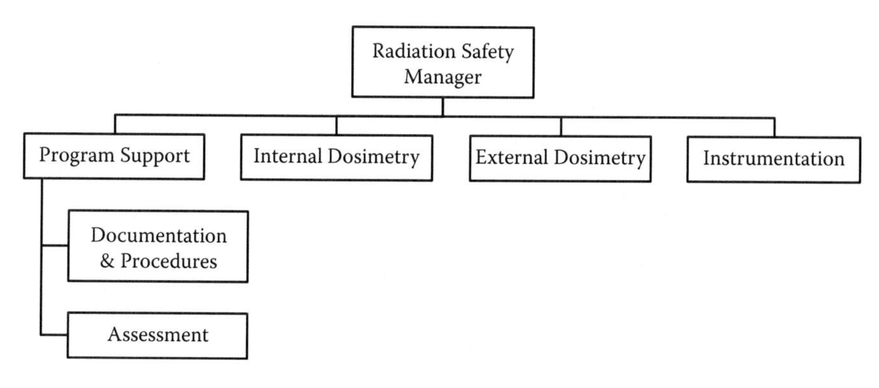

FIGURE 3.1
Radiation safety structure.

3.3 Roles and Responsibilities

The role of the radiological protection manager is one of leadership and authority and oftentimes reports to the environment safety and health manager. Oftentimes the radiological protection manager serves as a radiological control officer (RCO). Responsibilities of the radiological control manager include, but are not limited to, the following:

- Signature authority for all regulatory documentation.
- Ensure personnel are not overexposed/protected and the company is in compliance with regulatory requirements.
- Primary interface with management and regulatory agencies related to oversight.
- Confirms training is conducted in accordance with regulatory requirements and personnel are appropriately trained.
- Ensures adequate staffing to support company mission and schedule.
- Leads the organization in ensuring program and field execution is adequate to support the project schedule and operations and maintenance activities.

3.4 Radiological Program Flow-Down of Requirements

A company's radiological protection program is the basis for how a company ensures compliance with the regulations and defines program requirements. The general flow-down of requirements is depicted in Figure 3.2.

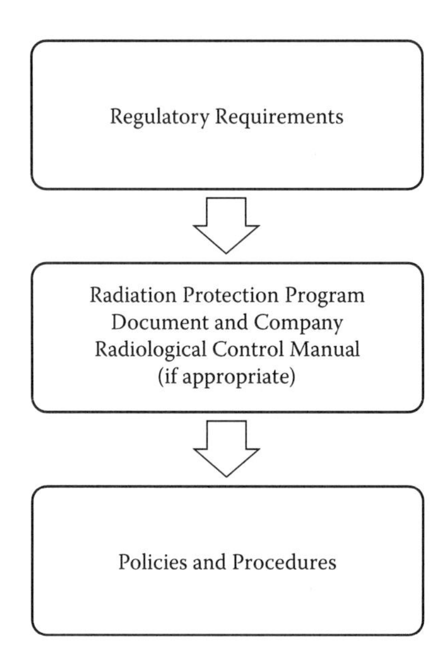

FIGURE 3.2
Radiological program requirements flow-down.

3.5 Regulatory Requirements

Both the NRC and DOE establish requirements and oversee the execution of radiological programs. Both regulatory agencies use Title 10 of the Code of Federal Regulations (CFR) to regulate. The NRC standards for radiation protection are listed in Title 10 CFR 20, and the DOE standards for occupational radiation protection are listed in Title 10 CFR 835. Both regulatory bodies have similar standards, and both the NRC and DOE have guidance documents for assisting program execution. The NRC issues a license to operate; the DOE approves initial start-up of a facility, but no formal license is issued—an authorization basis exists.

The NRC issues regulatory guides that provide guidance to licensees and applicants on implementing specific parts of the NRC's regulations, techniques used by the NRC staff in evaluating specific problems or postulated accidents, and the data needed by the staff in its review of applications for permits or licenses. The NRC also has developed standards associated with systems, equipment, and material. The DOE has several guidance documents for use in implementing provisions of 10 CFR 835, and numerous technical standards that promote radiological best practices and a consistent program and field approach for use in DOE facilities.

3.6 Radiation Protection Programs

Both the NRC and DOE require the establishment of a radiation protection program document that flows down regulatory requirements that are relevant to the licensee or company. Both NRC and DOE are similar in their approaches to radiation protection and require specific functional elements of the program to be defined. These include, but are not limited to, the following (depending upon the specific operations conducted):

- Management and organization
- Occupational dose limits
- Limits to the public
- Criteria for license termination (if operating under the NRC)
- Surveys and monitoring
- As low as reasonably achievable (ALARA) program
- Respiratory protection
- Records
- Reports
- Waste disposal (if operating under the NRC)
- Emergency provisions

The radiation protection program document, along with the license (if NRC), is legally enforceable. The radiological protection manager, operations manager and supervisors, and employees are held accountable to enforcement actions. The radiation protection program document also defines the functional elements of the program, from which radiological control manuals and policies and procedures are implemented. Most organizations/companies have subject matter experts (SMEs) assigned to assist in implementation of the functional elements in the programs and projects.

3.6.1 Radiological Control Manuals

As part of 10 CFR 835, the preamble stated that companies that have radiological control manuals, for enforcement purposes, would be granted presumptive compliance to 10 CFR 835. Many companies had radiological control manuals in place prior to rule issuance (DOE issued a radiological control manual prior to 10 CFR 835), so they continued to maintain the manuals after rule implementation. The radiological control manuals further define flow-down of regulatory requirements. Typical elements found in a comprehensive radiological control manual include the following:

- Standards of excellence in radiological control. Typically includes policy and commitment by management and leadership in radiological control, safety culture, communications, administrative terms, and applicability.
- Radiological standards. Typically includes internal and external dose limits, definition of administrative control levels, planned special exposures, and dose to minors.
- Conduct of radiological work. Outlines requirements for general planning of work and integration of ALARA into planning activities, radiological work permits, entry and exit requirements, survey and monitoring requirements, including frequency, respiratory protection and personal protective equipment, and definition of radiological areas.
- Radioactive materials. Identification, storage, and control of radioactive material. Labeling, packaging, and transportation are also addressed.
- Radiological health support operations. Defines practices and requirements associated with actual operational evolutions. Topics include dosimetry, both internal and external, control and standards for instrumentation (both portable and stationary), and management of contaminated personnel, including response to operational emergencies.
- Training and qualification. Defines the training and qualification requirements of personnel entering radiological areas, management, and radiological control personnel conducting work, including incorporation of a qualification standard if deemed appropriate for the operations.
- Records and reports. Outlines what records and reports are required to be maintained, the frequency for maintaining, and what is required to be reported to employees.

3.6.2 Radiological Protection Policies and Procedures

All companies and licensees have policies and procedures that define how radiation protection programs are implemented. They are the core of how a program is managed. Policies and procedures execute how work will be performed and flow down from the radiation protection program document, license, or radiological control manual. Each of the procedures executes some functional element of the radiation protection program. Table 3.1 identifies example policies and procedures that a company may use to implement each of its program functional elements and requirements.

In theory, if policies and procedures are executed as written, then personnel will be successfully protected from radiation, and operations and maintenance activities will be performed in compliance with regulatory

TABLE 3.1

Example Crosswalk to Program Functional Elements

Policy or Procedure	Functional Element
Commitment of management to radiological protection excellence policy	Administration
ALARA policy	Administration
Technical basis for air monitoring in the workplace	Administration
Development of radiological control technician procedures	Administration
Preparing radiological safety programmatic documents procedure	Administration
Internal dosimetry program procedure	Occupational dose limits
External dosimetry program procedure	Occupational dose limits
Area dosimetry procedure	Occupational dose limits
Tour dosimetry procedure	Limits to the public
Required radiological surveillance procedure	Surveys and monitoring
Job-specific workplace air sampling procedure	Surveys and monitoring
Real-time air monitoring procedure	Surveys and monitoring
Performance and documentation of radiological surveys procedure	Surveys and monitoring
Release surveys for tools, material, and equipment procedure	Surveys and monitoring
General sealed radiative source, material, standards, and radiation procedure	Surveys and monitoring
Application of ALARA in the design process procedure	ALARA
Radiological control technician and supervisor training program description procedure	Training
Qualification and training of radiological safety staff procedure	Training
Radiological safety instrumentation program procedure	Instrumentation
Dose investigation and documentation procedure	Occupational dose limits/records/report
Prospective dose assessment report procedure	Reports
Abnormal and unusual event control procedure	Emergency response

requirements. In addition, it is important to keep in mind that all documentation that is generated and used in the establishment and implementation of a program in the field is considered records and should be managed in accordance with regulatory requirements.

3.7 Radiological Protection Program Assessment

Several of the regulatory bodies require review of the functional program elements, by an independent person, within a required periodicity to ensure they are being properly executed and the program is continuing to improve. The independent person performing the assessment should not be directly associated with implementation of the program in the field.

One assessment option is to evaluate more than one functional/program element at a time so that the frequency of assessments is less; however, another option is to conduct such assessments on a monthly basis over a specified period of time (i.e., 3 years). There is no right or wrong approach when performing the assessment, but the rationale and logic should be documented and ensure all program elements are evaluated within their required periodicity as defined by the regulations. The radiation protection assessments are considered records and are often evaluated by the external agencies and are factored into the determination as to whether a program is sound. In particular, should an incident or event occur and a formal investigation is initiated, often the regulators will review results from past program and field assessments. The manager responsible for the radiological protection program is also responsible to ensure that not only is the program being effectively implemented, but also the program is independently evaluated and findings resolved.

3.8 Training in Radiation Protection

The level of training required for personnel performing work in or around radiological areas varies depending upon the level of risk of the work to be performed. The level of training required is commensurate with the level of potential radiological hazards to be encountered. For example, workers who do not directly come into contact with radiological material or areas would only be required to have general employee radiological training. However, workers who routinely work with radioactive sources may be required to have radiological worker I and II, source custodian training, and in some cases specific training for managing radioactive material. Radiological protection programs should have clearly defined training requirements associated with different types of workers. Table 3.2 is an example of a general radiation protection training crosswalk.

It is important to note that additional training is expected of radiation protection managers, radiological control officers, radiation protection supervisors, radiation protection technicians, and other disciplines that specifically work with radiological material, such as radiation-generating devices (RGDs).

The radiation protection manager is the final authority on all radiological decisions and is intimately involved with ensuring day-to-day operations are successfully executed. Typically, the radiation protection manager is involved with planning and decision making of work conducted and is responsible to oversee emergency response to abnormal situations. The radiation protection

TABLE 3.2

Example of a General Radiation Protection Training Crosswalk

Type of Radiological Training	Type of Worker
General employee radiological training	• Unescorted entry into controlled areas • Escorted or unescorted entry and may receive occupational exposure during controlled areas
Radiological worker I	• Entry into noncontaminated radiation areas or areas in which they are likely to receive doses exceeding 0.1 rem in a year • Work with sealed or fixed radioactive material that does not produce high radiation fields • Work with radiation-producing devices that do not produce high radiation fields (fields exceeding 0.1 rem in an hour)
Radiological worker II	• Entry is expected into high radiation areas • Entry is expected into contaminated areas • Work with unsealed quantities of radioactive material

manager may be responsible for not only program development and sustainment, but also successful execution of all radiation protection policies and procedures in the programs or projects. Because work is being conducted at the same time as the radiation protection program is being executed, it can be challenging to try to maintain compliance while still efficiently performing operations and maintenance activities. The radiation protection manager has the responsibility to ensure the company is protected and in compliance with regulatory requirements.

3.9 Radiation Protection Documentation

The radiation protection manager is primarily responsible for producing the documentation that demonstrates compliance with the regulatory drivers, but also demonstrates that personnel are not being overexposed to radiation and radioactive contamination. Documentation is generated at the start of each day, throughout the workday, and when work is finished at the end of the day. Instrumentation source checks, radiological surveys in the field, surveys conducted as part of a regulatory routine, work coverage surveys, sealed source surveys, and air monitoring analyses are all documented through radiological survey documentation and are considered records.

The quality of the documentation must be reproducible because the documentation is frequently relied upon to supplement dosimetry data in a determination of whether an employee was overexposed to radioactivity or radiation. In addition, the documentation is frequently reviewed by regulatory agencies to demonstrate compliance with the regulatory requirements. Many companies have procedures that address how to complete radiological surveys and data collection. Consistency in the manner by which documentation is completed also assists the radiological protection manager in ensuring all data presented is accurate and meets the intent of why the documentation or data was generated. Most companies today also use electronic media to document data and surveys, which further assists the radiological protection manager in ensuring current and accurate information.

3.10 Complacency in the Workplace

One of the greatest challenges of managing personnel in a nuclear environment is to ensure the workforce does not become complacent when executing procedures in the field. Over time, workers can become desensitized to hazards in the workplace, many that had been previously discussed, and tend to forget, overlook, or not believe an accident can happen if they perform a task a certain way. Radiological hazards are even more challenging to identify/remember and control because they visibly cannot be seen or the radiological worker has performed the task so frequently that he or she tends to minimize any hazards. The radiation protection manager should always be mindful of a culture that breeds complacency, in particular with those personnel who work with radioactive material on a daily basis.

3.11 Response to Abnormal Conditions

A significant role of the radiation protection manager is to manage and support emergency response actions related to radioactive material and accidents. Emergency response at nuclear facilities is of primary importance to both the NRC and the DOE. These agencies have increased their focus on how prepared companies are to handle emergencies to minimize potential impact in the event of unexpected conditions. Areas that are of particular concern, and expected increased regulatory oversight that should be managed, include the following:

- Prompt categorization and classification and implementation of protective actions and required notifications and reporting
- Demonstration that protective measures can and will be taken in the event of a radiological emergency

- Demonstration that emergency plans can be adequately demonstrated
- Adequate emergency planning procedures and training
- Regular drills and exercises
- Exercising their emergency plans with the regulatory agencies

Most companies are not aware that their emergency response protocol is not effective until after an accident has occurred, but this is one weakness where the radiation protection manager can always help improve performance and limit company liability.

3.12 Balance between Being a Company and a Worker Representative

The balance between meeting production goals and ensuring compliance with regulatory requirements is often a challenge for the radiological protection and ES&H managers. Many companies offer bonus or incentive programs for meeting production goals, which can, over time, lead to complacency in managing people and processes. It is prudent for a manager in the field to be mindful of business decisions that would favor production over worker protection.

Within both the nuclear safety profession and nonnuclear operations, there are several recent accidents that serve as stark reminders to us that without an adequate balance of safety and production, the mission and goals of the company cannot be achieved.

On April 20, 2010, the British Petroleum *Deepwater Horizon* spill occurred in the Gulf of Mexico. The accident involved a well integrity failure, followed by a loss of hydrostatic control of the well. This was followed by a failure to control the flow from the well with the equipment, which allowed the release and subsequent ignition of hydrocarbons. Ultimately, the emergency functions failed to seal the well after the initial explosion. Consequences of the accident resulted in the death of 11 people and devastation to the community, ocean, and aquatic life. The *Deepwater Horizon* accident is considered the largest accidental marine oil spill in the history of the petroleum industry.

Depending upon the reference, the number of causes of the accident varies, but one cause of the accident has been determined to be competing priorities between safety and production. The BP wells team leader was responsible for cost and schedule in addition to decisions affecting the integrity and safety of the well (and ultimately the workforce). It has been recognized that risks that are an inherent part of engineering processes in the petroleum and other industries were not fully recognized. Also recognized were the insufficient checks and balances for decisions regarding schedule to complete operational activities versus well safety. The final cost for cleanup of the *Deepwater Horizon* accident continues to grow, but most recently was valued in the several-billion-dollar range.

On March 11, 2011, a 9.0 magnitude earthquake struck Japan about 231 miles northeast of Tokyo off the coast of Honshu Island. The earthquake led to the automatic shutdown of 11 reactors at four sites along the northeast coast. Due to lack of diesel generators and off-site power to pump water into Fukushima Daiichi Units 1 through 4 to cool the nuclear fuel, as well as the hydrogen gas explosions inside the units, some of the nuclear fuel melted and led to radiation releases.

Although the earthquake initiated the tsunami, there are many professionals who believe the extent of the accident was "man-made," and that its direct causes were all foreseeable. There are a number of causal factors, including the fact that leadership failed to establish, implement, and maintain a culture that valued safety as a core value of business, and that there was insufficient independence of the regulatory agency overseeing the plant(s). As a result of the Fukushima accident over 16,000 people died from evacuation conditions and over 160,000 people (numbers vary) were evicted from their homes, and permanent exclusion zones have been established to ensure personnel are protected and to assist in the recovery effort.

A radiation protection or ES&H manager has the responsibility to emphasize a balance between safety and production. Although the current business environment tends to communicate "do more with less" and "do more even though risks may be higher," the radiation protection manager must understand that he or she does not have the authority to accept "operational necessity," and the consequences of action that justify risk or loss of life or equipment.

3.13 Summary

The study of radiation protection is the science and practice of protecting people and the environment from the harmful effects of ionizing radiation, and over the years the study of radiation protection has evolved. The primary role of the radiological protection manager is one of leadership and authority.

The regulatory agencies require the establishment of a radiation protection program document that flows down regulatory requirements that are relevant to the licensee or company. Both the NRC and the DOE are similar in their approaches to radiation protection and require specific functional elements of the program to be defined. The general flow-down of requirements is through the federal and state regulations, radiation protection program, and policies and procedures. In addition, there are regulatory requirements defined for a radiation protection program assessment, and training is a fundamental foundation for all radiological workers. The radiation protection manager should be the final authority on all radiological decisions and is intimately involved with ensuring day-to-day operations are successfully executed.

4

Worker Safety and Health

4.1 Introduction

Worker safety and health (WSH) is a multidisciplinary field that deals with the safety and health of the workers. The ultimate goal of a WSH program is to foster a safe and healthy workplace for workers as well as minimize potential negative impacts to the environment. An effective WSH program will have a strategy and a program that addresses the following:

- Supports the hiring and retention of skilled workers
- Ensures compliance to all regulatory and contractual requirements
- Ensures personnel understand how to anticipate, recognize, evaluate, and control workplace hazards
- Trains and educates workers
- Identifies improvement measures to support a safe work culture

A typical WSH structure is shown in Figure 4.1.

In a typical WSH organization structure the program/project support staffs represent the portion of the organization matrixed to programs or projects and provide direct support in the area of industrial hygiene and safety. The matrixed staff is organizationally direct-lined to the functional area to maintain independence. These staff members are heavily supported by the functional subject matter experts (SMEs) who are responsible for having expert knowledge in the health and safety regulatory requirements. This portion of the staff is integrated in and reports directly to the line in a matrix structure that also has responsibility to ensure work is safely performed and personnel are trained and qualified.

Training is an integral part of a worker safety and health program. In order to protect workers and the environment it is pertinent that the WSH professional be trained and knowledgeable in the hazard recognition, evaluation, and control process. WSH professionals should also possess a working knowledge of the regulatory requirements applicable to the company or project.

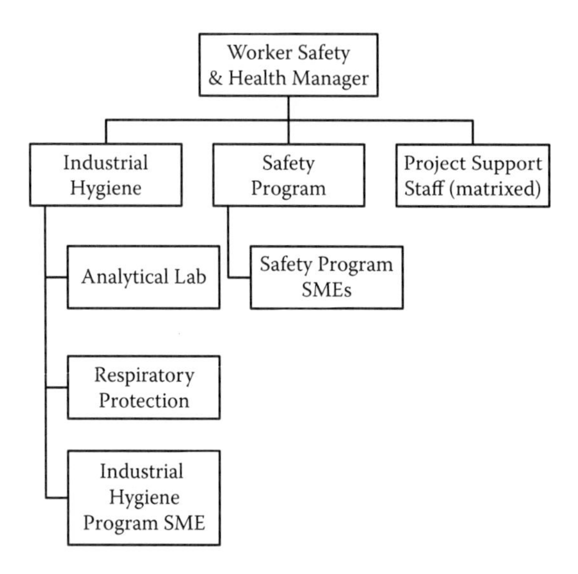

FIGURE 4.1
Worker safety and health structure.

The typical ES&H professional possesses degrees in the areas of science, engineering, statistics, and the medical field. Additional training may be necessary to become skilled as a WSH professional. Training will be discussed in further detail in Chapter 8.

When designing the worker safety and health program for your company, careful attention should be paid to the type of business and the business strategy, the size of the company, tasks being performed, and the regulations that are applicable to the scope of work to be performed. Once this information is known, then the WSH program can be developed and staffed.

Early on in this chapter it was mentioned that when building the WSH program, consideration should be given to the regulatory requirements that are applicable to the project scope and task. Below are some good sources of information that can aid in determining regulatory applicability:

- Occupational Safety and Health Act 29 CFR 1926
- Occupational Safety and Health Act 29 CFR 1910
- American National Standards Institute (ANSI)
- Occupational Safety and Health Administration (OSHA) website and letters of interpretations
- National Institute of Occupational Safety and Health (NIOSH) publications
- American Conference of Governmental Industrial Hygienists (ACGIH)
- National Electric Code (NEC)
- National Fire Protection Association (NFPA)

Other standards or publications may provide additional information; however, the above list represents the most widely used standards and publications. To aid in maintaining the level of knowledge needed to keep abreast of the regulatory changes that are applicable to the business, it is recommended that the many regulatory topical areas be distributed among the SMEs. Table 4.1 provides guidance on ways to distribute regulatory compliance responsibilities among the industrial hygiene SMEs. The table represents typical hazard areas that may be prevalent in a particular industry or business and is not considered to be comprehensive.

4.2 Industrial Hygiene Programs

Industrial hygiene (IH) is commonly defined as the science and art of anticipating, recognizing, evaluating, and controlling workplace hazards that may cause injuries and illness to workers. IH programs are designed to protect workers from exposure to chemical and physical hazards such as exposures to noise, thermal, and nonionizing radiation during the performance of tasks in the workplace. Industrial hygiene professionals are typically trained in the area of science, engineering, statistics, and technology. Knowledge in these areas is a key part of having the ability to anticipate, recognize, evaluate, and control workplace hazards. Some of the typical industrial hygiene programs that are listed in Table 4.1 should be developed and implemented along with applicable regulatory drivers. The table is not all-inclusive and the listing represents many of the prominent areas that are covered by a standard.

The typical areas referenced in the table are discussed in Sections 4.2.1 to 4.2.14. The information contained in these sections is designed to provide a starting point by which a program can be developed for the potential hazardous constituents. Reducing the risk of and minimizing overexposures can be achieved when a comprehensive program is in place and followed.

4.2.1 Asbestos

Asbestos is a naturally occurring mineral that is recognized for its insulating, heat resistance, and tensile strength properties. Historically, asbestos has been used for everything from fireproofing, woven into fabrics, and mixed into cement. There are thousands of products that may be identified as asbestos-containing materials (ACMs). Some of the products are listed in Table 4.2.

The primary health hazards caused from exposure to asbestos are asbestosis and mesothelioma. Asbestosis is an inflammatory condition of the lungs that can cause shortness of breath, coughing, and eventually scarring of the lungs that makes it hard to breathe. Mesothelioma is a rare cancer that affects

TABLE 4.1

Example Industrial Hygiene Programs

Subject/Topic	Regulatory Drivers and Guidelines
Asbestos	29 CFR 1910.1001
	29 CFR 1926.1101
Beryllium	10 CFR 850
	29 CFR 1910
Biosafety	Centers for Disease Control and Prevention
Chemical safety (hazard communication)	29 CFR 1910.1200
Chemical hygiene	29 CFR 1910.1450
Confined space	29 CFR 1910.146
	29 CFR 1915
Food safety	Food and Drug Administration
	29 CFR 1910.141(h)
	19 CFR 1926.51(d)
	Centers for Disease Control and Prevention
Heat stress	29 CFR 1910 (general duty clause and other applicable standards)
	Centers for Disease Control and Prevention
Lasers	ANSI 136
Lead	29 CFR 1910.1025
	29 CFR 1926.62
	OSHA permissible exposure limit
	ACGIH threshold limit value
Nanotechnology	29 CFR 1910 (general duty clause and other applicable standards)
Noise	29 CFR 1910.95
Personal protective equipment	29 CFR 1915.152
	ANSI Z87.1-1968
Refractory ceramic fibers (RCFs)	29 CFR 1910 (general duty clause and other applicable standards)
Respiratory protection	29 CFR 1910.134
	ANSI Z88.2
Silica	29 CFR 1910.1000

the lining of the lungs, chest cavity, or abdomen. Careful planning must be made when work is to be performed using ACMs.

The OSHA has outlined in 29 CFR 1910.1001 and 29 CFR 1926.1101 specific controls, training for workers, and sampling protocols for work performed with asbestos and asbestos materials. In order to control exposures to ACMs close adherence to the OSHA standard is necessary. A comprehensive asbestos program consists of training for supervisors and workers, medical monitoring, engineering control, personal protective equipment, and air sampling (personal, area, and clearance).

TABLE 4.2

Asbestos-Containing Materials and Products

Elevator car brake shoes	Elevator equipment panels
Brake and clutch pads	Ceiling tiles and texture
Siding	Wall panels
Cloth wire insulation	Electrical panels
Boiler insulation	Cooling towers
Heat shields	Tank insulation
Floor tile mastic	Heat-resistant gloves
Brake linings	Shingles
Gasket	Carpet mastic
Adhesives	Appliance components
Acoustical plaster	Chimney fluke lining
Chalk boards	Electrical breakers
Fire blankets and curtains	Fire doors
Duct insulation	Furnace insulation
Elevator equipment panels	Asphalt floor tiles
Paint and coatings	Roofing products
Insulation cloth	Fire proofing insulation

4.2.2 Beryllium

Beryllium is a metal widely used in industry because of its properties, such as light weight, high tensile strength, and good conductivity. Industrial processes where work is being performed with beryllium and can lead to levels above the occupation exposure limits are listed in Table 4.3. The primary health hazards caused from exposure to beryllium are skin sensitization and beryllium sensitization, which can lead to chronic beryllium disease, a disease that can affect different tissues and organs.

TABLE 4.3

Occupations That May Involve Exposure to Beryllium

Sporting goods	Nuclear applications
Electronics	Machining
Die casting	Production of beryllium metals
Molding of plastics	Thermal casting
Rocket parts	X-ray tube window manufacturing
Computer component manufacturing	Handling and assembling of beryllium parts
Manufacturing of ceramic	Smelting and refining (beryllium)
Laboratory work involving beryllium	Welding electrodes
Manufacturing of navigation systems	Manufacturing dental plates
Heat shields	Decontamination and decommissioning work

Beryllium is also a known cancer-causing substance, and exposure to high levels can cause lung cancer. The OSHA has specific requirements for working with beryllium. In addition, if work is being conducted at a Department of Energy site or project, 10 CFR 850 is applicable and outlines specific requirements for working with beryllium, training, beryllium registry participation, and surveillance for workers. Before developing a beryllium program a determination must be made of which of the two standards are applicable.

4.2.3 Biosafety

Biosafety requirements are written to ensure that workers understand the hazards of the biological agents and the means by which to protect themselves from potential exposure. Biological agents are generally described as a type of bacterium, virus, protozoan, parasite, or fungus. A good example of a common application of biosafety is management of blood when responding to an accident or emergency. Biological agents may also be used as a warfare agent or are often used for research purposes. Health effects from exposure to biological agents can range from mild symptoms to death depending on the agent and duration of exposure.

Biosafety regulations are published by the Centers for Disease Control and Prevention (CDC). Generally, companies that work with potentially harmful biological material will have a committee that is responsible for making decisions regarding biosafety and administering the program. The biosafety committee is also responsible for monitoring compliance with applicable standards. A responsible individual (RI) is appointed generally to ensure compliance with regulatory requirements. The RI is also responsible for program development to include assisting with the development of training for the worker.

4.2.4 Chemical Safety and Hygiene (Hazard Communication)

Chemicals in the workplace can pose a wide range of health hazards to various body parts. Health hazards can include irritation, sensitization, carcinogenicity, and death. In addition, some chemicals can exhibit physical hazards, such as flammability, corrosion, and reactivity. The OSHA Hazard Communication Standard is designed to ensure that employees have information about these hazards and protective measures to reduce the risk of overexposures. OSHA requires manufacturers to develop safety data sheets (SDSs) to communicate chemical components and the associated hazards of products.

All employers are required to have a written hazard communication program if hazardous chemicals are present in the workplace, and provide labeling of chemicals and training for employees on how to safely work with the chemicals. A chemical hygiene plan is required by OSHA for laboratories working with hazardous chemicals.

4.2.5 Confined Space

Confined spaces are areas or spaces with limited means of entry or exit and that are not designed for continuous human occupancy. Many workers die annually as a result of working in confined spaces. These spaces are large enough for workers to enter and perform various types of jobs. Examples of confined spaces include pipes, tanks, vessels, tunnels, silos, storage bins, hoppers, vaults, pits, manholes, and duct work.

OSHA has specific requirements for employers to follow in order to allow employees to work safely in these spaces. A confined space program must also include an inventory, training for workers, posting, and monitoring of these spaces to ensure spaces are safe for workers to enter and perform work. A solid confined space program is critical to worker safety and health when working in confined spaces.

4.2.6 Food Safety

Food safety refers to the handling, preparation, and storage of food items. Food that is not appropriately stored or prepared can result in food-borne illnesses that can be a severe public health issue, not to mention the financial liability to the company. Diseases and pathogens can be transferred from food to humans. Food safety is covered primarily by the U.S. Department of Agriculture and the Food and Drug Administration. However, OSHA can enforce food safety through its general duty clause. Employers are expected to have instituted and enforce work practices to ensure food items are safe for consumption. This includes proper handling, storage, and preparing of food items.

4.2.7 Heat Stress

Each year many workers are exposed to excessive heat in the workplace both indoors and outdoors. Some of these workers eventually will die as a result of their exposure. Work environments involving high ambient air temperatures, radiant heat sources, direct physical contact with hot objects such as furnaces, or strenuous physical activities can potentially cause heat-related illness. The employer is charged with the responsibility of protecting employees while working in these environments. A practical heat stress program begins with a heat stress management plan that is supported by management and includes providing training of heat stress protective measures, signs, and symptoms, and monitoring (medical and atmospheric).

4.2.8 Lead

Exposure to lead can occur in operations such as welding, torch cutting, sanding, abrasive blasting, grinding, batteries, demolition activities, firing ranges—ammunition, and removal of lead-based paint. Health effects of exposure to lead include lead poisoning and in some cases death. When working with lead the employer's responsibilities include not only the worker, but also the environment. Lead that is emitted to the environment (for example, through soil, water, and air) can adversely impact the health of the public. When working with lead a sound program must be implemented to prevent exposures. The program should include at a minimum control mechanisms (e.g., engineering, work practices, personal protective equipment (PPE)), exposure monitoring, training, and medical surveillance for workers.

4.2.9 Nanotechnology

Nanotechnology is the study and application of extremely small particles. Nanotechnology involves the understanding, manipulation, and control of matter at dimensions of roughly 1 to 100 nanometers, which is near-atomic scale, to produce new materials, devices, and structures. The size of nanoparticles can render them extremely hazardous to workers when exposed. Therefore, a proper hazard analysis is necessary prior to performing tasks involving nanoparticles. In addition, air monitoring and medical monitoring (where applicable) should be included in development and implementation of a nanoparticle safety program.

4.2.10 Hearing Conservation—Noise

Every year millions of people are occupationally exposed to noise. Noise in the workplace can have a lasting effect on the worker, creating hearing loss that cannot be repaired. A hearing conservation program is required by OSHA and is necessary in order to ensure hearing conservation is integrated into the injury and illness prevention program. A hearing conservation program will consist of monitoring, engineering controls, audiometric testing, hearing protection devices, employee training, and record keeping. It is important to evaluate the adequacy of the program and make improvements as needed. One important indication of whether the program is effective is a noted reduction in hearing loss cases.

4.2.11 LASER

The acronym LASER stands for light amplification stimulated emission of radiation. Lasers produce an intense and highly directional beam that can penetrate and cause tissue damage. Laser usage includes medical applications, welding and cutting, laser nuclear fusion, surveying, communication, heat treatment, and spectroscopy. The primary health effect of laser use is the possibility of eye injury and, to a lesser degree, skin injury. Unprotected laser exposure to the eyes can result in eye injuries such as burns to the retina, cataracts, and blindness. Therefore, when working with lasers the appropriate control mechanisms must be in place to ensure protection of the eyes. These control mechanisms should include having the appropriate interlocks and eye protection for the worker. The employer is required to designate a laser safety officer (LSO) when working with Class IIIB and IV lasers. The employer is also required to ensure that the LSO is trained and knowledgeable to function in the role of LSO. The LSO is responsible for developing the laser safety program on behalf of the employer. A laser safety program should include operating procedures with hazard controls and safety measures, training for employees that work with lasers, personal protective equipment, medical surveillance, and a process to investigate incidents when they occur.

4.2.12 Personal Protective Equipment

Personal protective equipment (PPE) is a hazard control method used in helping to prevent employee exposures to various chemicals or chemical compounds and physical hazards. The use of PPE to prevent exposures is the least preferred hazard control option since the use of PPE can potentially add additional stress to the worker, and therefore become a hazard in certain cases. Before prescribing and using PPE, a thorough hazard assessment must be completed to determine the type of PPE that would be most effective for the hazards. Prior to distributing PPE to the employee, some level of training must be provided so that he or she understands how to use the PPE properly. Improper usage of PPE may not provide the level of protection desired or required. It is important to make use of the many PPE selection guides that are available to assist in making the right selection choice. The ANSI standard and the OSHA regulations for PPE provide the requirements for a compliant PPE program.

4.2.13 Refractory Ceramic Fibers

Refractory ceramic fiber (RCF) materials are generally treated similar to asbestos-containing material by many as a best management practice. Lacking an OSHA specific standard, RCFs can be regulated under the OSHA general duty clause. Many occupational health professionals are concerned that RCFs can be just as hazardous when inhaled as asbestos-containing materials. RCFs are typically used in applications requiring high-heat insulation, such as furnaces. It is common to use the program and policies for RCFs that are in place for asbestos-containing material. If an asbestos management program is not in place, an RCF safety strategy should be developed using the standard for asbestos as a guide.

4.2.14 Respiratory Protection

Respiratory protection devices are another form of personal protective equipment (PPE) used in the workplace. There are many different types of respiratory protection devices that are available to provide protection against the many inhalation hazards. Respirators can protect against a variety of hazards, such as oxygen-deficient environments, dusts, mists, gases, and vapors. A comprehensive respiratory protection program encompasses training on use, selection and maintenance, medical surveillance, and fit testing. ANSI and OSHA should be consulted for the requirements for an appropriate and compliant respiratory protection program. Remember, respiratory protection is the least desired hazard control method to use for worker protection.

4.2.15 Silica

Exposure to crystalline silica in the workplace often occurs during tasks involving cutting, sawing, drilling, and crushing of concrete, brick, and products made of stone. Operations using sand-based products such as sand blasting, manufacturing glass, masonry work, cement and granite cutting, demolition, and work in foundries can result in worker inhalation of silica particles that have been dispersed in the air. Crystalline silica is classified as a carcinogen, which means that inhalation of silica can cause cancer in humans. Inhalation of silica can cause damage to the lungs and lead to silicosis or even lung cancer. A comprehensive silica exposure control program should include a comprehensive hazard analysis of tasks, engineering control such as ventilation, personal monitoring, and PPE, to include respiratory protection equipment if warranted.

4.3 Occupational and Industrial Safety

Industrial safety is the part of the WSH program that focuses more on the physical hazards as opposed to chemical hazards, as seen in the area of industrial hygiene. We will cover some pertinent areas on safety that will come into play when designing a comprehensive safety program. Table 4.4 lists a number of industrial safety programs that may be part of the company's program. This list is not comprehensive and is only intended to get the manager started in formulating program development.

4.3.1 Aerial Lifts

According to OSHA, aerial lifts are vehicle-mounted, boom-supported aerial platforms, such as cherry pickers or bucket trucks, used to access utility

TABLE 4.4

Example Industrial Safety Programs

Subject/Topic	Regulatory Drivers and Guidelines
Aerial lifts	29 CFR 1926.453
Construction safety	29 CFR 1926
Cranes/hoisting/rigging	29 CFR 1926.753
Electrical safety	29 CFR 1910 Subpart S
	29 CFR 1926 Subpart K
Elevators and escalators	29 CFR 1917.116
	29 CFR 1926.552
Ergonomics	29 CFR 1910.5
Explosives	DOE M 440.1-1
	29 CFR 1910.109
	29 CFR 1926.904
	29 CFR 1926.905
	29 CFR 1926.900
Fall protection	29 CFR 1926 Subpart M
Firearms	27 CFR 478
Powered industrial trucks (fork trucks)	29 CFR 1910.178
Ladders	29 CFR 1926.1053
Lighting—illumination	29 CFR 1926.56
	29 CFR 1915.85
Lockout, tagout	29 CFR 1910.147
Machine guarding	29 CFR 1910 Subpart O
Pressure safety (including compressed gases, cryogens)	OSHA Technical Manual Section IV: Chapter 3, 29 CFR 1910, 29 CFR 1915, 29 CFR 1626 (various subparts)
Signs	29 CFR 1926 Subpart G
	May be specific to hazard
Traffic safety	See national and state transportation regulations
Walking-working surfaces	29 CFR 1910 Subpart D
Welding, cutting, brazing	29 CFR 1910.252

lines and other aboveground elevated job sites. The major causes of fatalities when using aerial lifts include falls, electrocutions, and collapses or tip-overs. Aerial lifts have replaced ladders and scaffolding on some jobs for many applications. Employers must take measures to ensure the safe use of aerial lifts by their workers if they are required to use them in the course of their employment, including appropriate training and demonstration of skill. Aerial lifts include, but are not limited to, the following:

- Aerial type ladders
- Boom platforms (expandable and joining)
- Vertical towers

There are many hazards associated with aerial lifts; therefore, it is incumbent upon the employer to conduct a proper hazard assessment prior to using these lifts. The employer is required to provide training on safe usage, and implement the proper controls to ensure safety of the worker.

4.3.2 Construction Safety

The area of construction safety covers all projects or tasks that must be performed using the OSHA construction safety standards. It is important for the employer to understand the difference between a task or project that is subject to the OSHA general industry standard (29 CFR 1910) and the OSHA construction standard (29 CFR 1926). Currently these standards overlap in some areas and are not always in alignment with the specified requirements. To eliminate confusion to workers, management should ensure that policies, procedures, and training be in alignment with 29 CFR 1926 for construction work activities when it applies.

4.3.3 Hoisting and Rigging

Hoisting and rigging safety for purposes of this discussion will include crane safety. Lifting devices are used by many employees daily to lift equipment and supplies that are too large to be manually handled. Oftentimes lifting is performed using cranes and other equipment used to restrain or secure the material being lifted. These equipment and devices must be stored properly and inspected to ensure that they are safe for their intended use. In addition, the equipment operator must be trained on the operational aspect of each type of equipment used.

4.3.4 Hazardous Energy

Working with hazardous energy sources can be dangerous even for the skilled and well-trained worker. These hazards can cause burns, shocks, electrocution, blunt force trauma, and death. OSHA has written safety requirements designed to protect employees that can potentially be exposed to hazardous energy. These hazards can include the release of stored energy, electrical hazards, electrocution, fires, and explosions in the workplace. Electrical hazards are also addressed in specific standards, such as shipyard employment and marine terminals. Workers that can be exposed to hazardous energy during the performance of their jobs include engineers, electricians, operators, mechanics, and technicians. Office workers and salespeople can also be exposed indirectly to hazardous energies. Employers are required to have a program in place to protect workers from these hazards, in which training is an integral part.

4.3.4.1 Lockout, Tagout

Lockout, tagout is a control method used to prevent exposures to hazardous energy. The lockout, tagout program and process is designed to protect workers while servicing equipment to prevent the release of hazardous energy during the performance of maintenance activities. An effective program includes the following elements at a minimum: a documented and hazardous energy control program and training of employees on the requirements of that program. The training program must cover the attributes of the employer's hazardous energy control program, elements of the program that are relevant to the employee's work assignment, and the requirements to Title 29 CFR Part 1910.147 and Part 1910.333 where applicable. It should be noted that the term *hazardous energy* encompasses not just electricity, but all other sources of stored energy that may be in a piece of equipment. Therefore, it is important for the ES&H professionals to understand where energy may be potentially stored and released during operations and maintenance activities.

4.3.5 Elevators and Escalators

More and more multilevel buildings are designed and built with elevators that are used to provide access to the many floors. As a result, elevators and escalators are widely installed in many buildings. Consequently, elevators are required to be inspected and the documentation of that inspection readily accessible. According to the OSHA, elevators and escalators must be inspected at specified intervals, inspection must be conducted by designated workers with the appropriate knowledge and skills, and the result of the inspection posted in the vicinity of the elevators or escalators.

4.3.6 Ergonomics

Ergonomic-related injuries can be easily classified as one of the leading work-related injuries occurring today. A large segment of the population has jobs that require them to perform tasks that are repetitive in nature. These types of tasks, if not properly designed, taking into consideration the body posture and without the appropriate work-rest regimen, can result in serious injuries. In addition, when evaluating the resulting injuries, it is sometimes not clear if the injury happened at work or at home, especially if the injury was not reported at the onset of discovery. Sometimes it is difficult to distinguish between ergonomic injuries that occur on the job and the injuries that are a result of activities performed at home. Therefore, when designing the ergonomic program for your company, careful consideration should be given to the structure of the program and process used for determining what ergonomic-related injury is work related. An effective ergonomic program has a prevention element, training, and means to treat injuries when they occur.

4.3.7 Explosives and Blasting Agent Safety

Working with explosive chemicals and devices is not a trivial task. Safe work practices are necessary in operations involving the development, testing, handling, and processing of explosives or assemblies containing explosives. There are standards and requirements for working with and handling explosive materials that should be used in designing a program for worker safety. Not only is the handling of explosive material important, but it is also pertinent to ensure that it is stored appropriately to avoid detonation. Several organizations have published instructions that can be used in designing and implementing an effective explosive safety program.

4.3.8 Fall Protection

Falls are one of the most common work-related injuries experienced on the job. Many fall-related accidents result in serious injuries or death. In fact, falls are viewed as being among the leading cause of death in the construction industry. OSHA requires that employers have the necessary controls in the workplace to prevent fall-related injuries. Fall protection systems generally will fall into one of two categories—fall arrest and fall restrain; both are designed to prevent a worker from falling. Regardless of the system chosen, employers have the responsibility to provide the appropriate system for the job to be performed, and to train workers on the selection, use, and maintenance of the fall protection systems.

4.3.9 Powered Industrial Trucks

Powered industrial trucks, commonly referred to as fork trucks, are used by many companies to lift and transport bulk supplies. Any mobile power truck that can be propelled and is used to carry, push, pull, lift, or stack materials is classified as a powered industrial truck.

In addition, powered industrial trucks can be ridden or controlled by an operator walking alongside of them. The OSHA standard has specific requirements for employers, to include development of a training program, ensuring that trained operators know how to use the equipment safely, which is demonstrated through worker evaluation, certifying that each operator has received the required training, and developing and administering refresher training at least once every 3 years or whenever an operator demonstrates a deficiency in the safe operation of the truck.

Prior to purchasing or using powered industrial trucks in the workplace it is incumbent upon management to ensure that the proper program is in place to ensure worker safety. Consulting the OSHA regulation is a good place to start in designing a safe and compliant program.

4.3.10 Ladders

Ladders are found in every home and workplace, and OSHA publishes a variety of fact sheets on safe usage. Since ladders come in various sizes and can be constructed out of different materials, it is important to ensure that the right type of ladder is selected for the task. For example, one would not want to choose a ladder constructed of metal if working around live electrical components. In such case, the use of a ladder can present additional hazards to the worker. Ladders can be constructed of fiberglass, wood, or metal. As with all equipment, ladders should be inspected prior to usage and the user must ensure that it is the proper length to reach the height needed to avoid the temptation of standing on the top rung. Providing the right tool for the job is the responsibility of management, and in many cases the use of a ladder is the right tool for the job.

4.3.11 Lighting—Illumination

The importance of having adequate lighting in the workplace is often underestimated. Employers are required to provide adequate lighting in walkways and work areas to ensure employee safety. The needed quantity of light for any particular space is dependent upon the work that is being done and the environment in which the work is being performed. Benefits of good lighting in the workplace include improved accuracy, reduced risk of injuries and illness, and overall improvement in the performance of work. Before designing the lighting plan or making changes to workplace lighting it is a good idea to consult the workers for their input. Adequate lighting can be a critical element to improve productivity, safety, and security.

4.3.12 Machine Guarding

Moving equipment parts have the potential to cause severe workplace injuries, to include crushed fingers or hands and loss of body parts. Machine guarding is an essential element in protecting workers from these types of injuries. Any machine part, function, or process that can cause injury must be safeguarded with the appropriate type of guard. Various standards are written to address machine guarding to aid in the prevention of injuries. These hazards are addressed in the OSHA general industry and construction, longshoring, and marine terminals standards. It is important to consult the standard that is applicable to the industry in which work is performed.

4.3.13 Pressure Safety

A pressure vessel is considered a tank or vessel that has been designed to operate at pressures above 15 psig. Cracked and damaged vessels can result in leakage or rupture failures creating potential health and safety hazards of leaking vessels that can include fires and explosion hazards. If a pressurized vessel fails it can injure or kill people in the vicinity of the equipment as well as cause damage to property. Therefore, it is paramount that pressure vessels be designed, installed, and operated in accordance with the appropriate standards.

4.3.14 Signs

The use of signs is an integral attribute to a safety program. Signs are used as a means to inform workers of the condition of an area, equipment, or process. OSHA has specified certain conditions for when the use of signs is appropriate, as well as the type of sign. Signage is intended to warn of workplace hazards and instruct employees on the proper precautions to take to avoid injuries. When signs are used in the workplace, it is necessary to ensure that the workers have an understanding of the meaning of each sign and the appropriate action to take when the sign is present.

4.3.15 Traffic Safety

A traffic control plan is necessary in areas that have vehicle movements and workers commingled. These areas should have a well-thought-out traffic control plan, to include traffic control devices and signs. A good traffic safety plan for the workplace should include the following.

- Signs—include information of speed and condition or status of work zones.
- Work zone protection devices—such as concrete barriers to limit the feasibility of motorists entering the work area.
- Traffic control identifiers—directing the flow of traffic away from areas occupied by people.
- Flaggers—providing temporary control of traffic. Note: Flaggers should be illuminated and trained on the use of proper signaling methods.

4.3.16 Walking-Working Surfaces

Slips, trips, and falls constitute a large majority of the accidents occurring in the workplace, resulting in the majority of workplace injuries. Poor housekeeping accounts for a large percentage of slips, trips, and falls within the work area. Work practices that include good housekeeping can be a great way to reduce these types of injuries. The OSHA standard for walking-working surfaces outlines the requirements for employers to provide a work environment that should minimize accidents and injuries.

4.3.17 Welding, Cutting, Soldering, and Brazing

Employers are expected to provide specific training, safe work procedures, and protective equipment to employees who perform tasks that involve welding, cutting, soldering, and brazing. Welding, soldering, and brazing are used to join two pieces of metal together, whereas cutting is used in separating metal. Fumes can be generated during the performance of these processes; therefore, conducting a hazard analysis before beginning tasks involving these processes is necessary to ensure protection of the worker.

4.4 Medical Surveillance Program

Medical surveillance is an integral part of the worker safety and health program. Medical surveillance is the systematic assessment of employees exposed or who could potentially be exposed to hazards in the workplace. A comprehensive medical surveillance program is designed to monitor workers for adverse health effects and determines the effectiveness of exposure prevention and control strategies. The goal of the medical surveillance program is to prevent occupational illness and injuries in the workplace. A good medical surveillance program requires close coordination with the occupational health physician and the safety and health manager and professionals. The OSHA and ACIGH standards should be consulted to determine if a particular chemical or material has a medical surveillance requirement to further ensure protection of employees. The information gained through the reviews of the two standards is necessary in determining the elements of the medical surveillance program. The occupational health provider's role in implementing the medical surveillance program is covered in more detail in Chapter 5.

4.5 Occupational Exposure Limits

In order to protect workers from experiencing adverse effects from their employment, the OSHA sets enforceable permissible exposure limits (PELs) to protect workers (29 CFR 1910 Subpart Z and 29 CFR 1926 Subpart Z). These limits are enforceable regulatory levels of a substance that can be present in the air during the course of performing work-related tasks in which most workers can be exposed to without experiencing any adverse effects. The PELs are based on an 8-hour workday time-weighted average (TWA). In the event the work shift exceeds 8 hours, adjustment to the 8-hour TWA must be made to ensure compliance with the PELs.

The WSH program is tasked with designing programs that are supportive of maintaining employee exposure as low as reasonably achievable and well below the occupational exposure limit (OEL). When developing the WSH program, the substances that will be utilized or produced during operation should be compared against the OEL to determine applicability. The OELs are published by various agencies, to include OSHA and the American Conference for Governmental Industrial Hygienists (ACGIH). These agencies are the ones that are most commonly used for worker safety and compliance purposes. The OELs that are published by OSHA are legally enforceable.

4.6 Analytical Laboratory

A reputable analytical laboratory is necessary to complement a worker safety and health program. Laboratory analysis can be performed internally or externally through a contractual agreement with a laboratory service provider capable of performing the analytical needs of the organization. To ensure samples are analyzed based on the appropriate and required procedures, the laboratory, whether it is internal or external, should be certified by the American Board of Industrial Hygiene (ABIH) laboratory accreditation process. Laboratory certification adds credibility to the sample results, and provides some assurance that the lab performing the analyses has some rigor in the process to ensure accuracy of data. Depending upon the contaminant, a certified laboratory may be required by regulations.

4.7 Summary

The profession of worker safety and health is a very diverse field that functions at the core of the culture designed to protect workers from hazards in the workplace. We have covered briefly some of the most prominent areas of a worker safety and health program that a manager must consider when developing an effective WSH program. The areas that were covered are not intended to be inclusive. However, they are intended to serve as an indication of the importance of evaluating each of the programs, processes, and tasks that a worker will be involved in to ensure that the regulatory requirements are adhered to and the safety of workers is at the forefront of the decision-making process.

Recognizing that the field is broad and widely governed by regulations and standards, there is a need to engage a professional staff with the appropriate skill level. A rule of thumb to consider when pulling together the ES&H staff is not all ES&H professionals and position requirements are equally created. Workers have rights and deserve a safe workplace. Employers and managers are responsible and are held accountable for providing workers a safe workplace.

5

Occupational Health and Medicine

5.1 Introduction

For many companies and businesses the management of the occupational health group, or some facet of an occupational health function, organizationally resides within the ES&H department. This chapter will focus on the employer's responsibility for providing occupational health monitoring of workers and the risk associated with workplace injury and illness.

Occupational health refers to the identification, control, and management of risks that are associated with chemical, biological, and physical hazards in the work environment. Safety and health professionals (including those related to other disciplines of ES&H) manage worker risks from the immediate work environment versus health care professionals who assess and medically treat the worker, whether it be from prevention of illness and injury and treatment beyond first aid.

The primary regulatory driver for managing occupational health requirements related to commercial businesses, and those government agencies regulated under the Occupational Safety and Health Administration (OSHA), is within the 29 CFR 1900 series. Within the Department of Energy (DOE), occupational health requirements are found within 10 CFR 851. When managing an occupational health organization, the ES&H manager must be aware that there are several facets of the program that will require his or her attention and leadership skills:

- General management of medical staff
- Execution of medical functions
- Prevention and wellness programs
- Injury case management and reporting

It is important to note that the ES&H manager does not necessarily need to be knowledgeable about the medical profession, but rather he or she needs to rely on the medical professionals who understand the medical requirements associated with the occupational health program and feedback from the medical professionals when making company decisions.

5.2 Management and Administration

The size and organizational structure of the occupational health group will vary depending upon the type of operations being conducted (level of health risk) and the size of the company. Small companies may subcontract their occupational health services, whereas larger companies may directly employ a doctor, nurse, medical technicians, and administrative staff. In all instances, the occupational health physician is the primary caregiver and has responsibility for ensuring compliance with medical requirements and final signature authority. Although other individuals may treat a patient, ultimately it is the physician that must concur and have final signature authority on all treatment.

A typical organization chart for a health service organization is shown in Figure 5.1. It is important for the ES&H manager to work with the physician in communicating roles and responsibilities and expectations of the medical staff. Given the type of work the medical staff perform, a certain level of authority and trust is given to them by the patient and the company; authority and trust are inherent with the job they perform. Consequently, the ES&H manager, together with the physician, should clearly define roles and responsibilities for each medical staff position, communicate performance expectations, including patient management, and then enforce accountability of their individual performance to the defined expectations and requirements.

One of the greatest challenges the ES&H manager will face when managing the occupational health staff is integrating the two different work environments, personalities, and their experience and training.

Medical staff are typically trained in a hospital or medical office environment, not in a typical industrial work environment. Roles and responsibilities

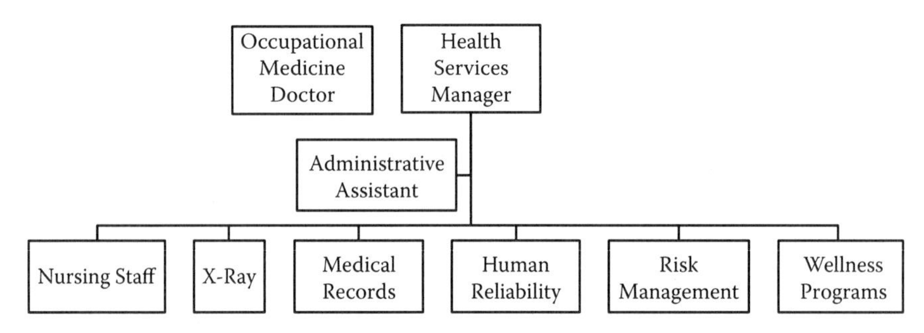

FIGURE 5.1
Occupational medicine organization chart.

and expectations of performance are different in a medical setting versus an industrial business setting. The primary goal of the medical setting is patient treatment and comfort. The primary goal of a business, in addition to the health and safety of the workforce, is successfully meeting production goals, obtaining a profit, and growth of the company. The two business models do not originate from similar goals and objectives; therefore, it is the ES&H manager's responsibility to work with the medical staff to ensure there is a common alignment of goals and objectives for an efficient and effective work environment.

A part of the alignment is the need for the ES&H manager and physician to communicate how the medical staff contribute to the overall company success. Understanding how every person contributes to the success of the company is important because the majority of the medical staff are not accustomed to performing their work in an industrial setting. Management of the medical staff should address not just patient management, but also day-to-day operations of the clinic itself.

On a daily basis there are routines that must be performed to ensure medications and medical standards are being followed, records management is being conducted, and general housekeeping of the clinic. There are a number of regulatory requirements associated with medications, medical supplies, and overall work conditions of the clinic, and it is ultimately the responsibility of the ES&H manager to ensure the clinic is being appropriately managed. In addition, any changes to policies, procedures, goals, and objectives need to be effectively communicated and reinforced.

One of the most important areas of management and administration of medical staff is ensuring all personnel are up-to-date with training and necessary qualifications. It is the ES&H manager's responsibility to ensure workers are not being treated by unqualified personnel; therefore, it is recommended a training tracking database be established that tracks required training classes, by position, for each member of the medical staff.

The training tracking database should include a process that notifies medical personnel of when their required training expires within 90 days of training expiration. It is also the responsibility of the ES&H manager to ensure funding and time are allocated to pay for personnel to maintain their qualifications.

5.3 Functions of the Occupational Health Staff

There are a number of functions the medical staff perform as part of an occupational health program. Below is a discussion of several common functions. These functions are performed by all different members of the staff, but ultimately are overseen by the physician and the ES&H manager.

5.3.1 Preemployment Physicals

Preemployment physicals are performed on prospective employees to verify they are physically capable to perform the job which they were hired to do. All physical requirements associated with each job should be documented, up-to-date, and understood by the medical professional performing the exam. Preemployment physicals generally consist of the following (this will vary depending upon the work being conducted and company policy):

- Medical history
- Vital signs and physical exam (may be tailored to type of work to be performed)
- Vision and hearing testing
- Drug and alcohol testing
- Job-specific tests such as blood work and respiratory

Almost all of the tests are designed to determine whether the future employee will be able to perform the job, but also the tests will establish baselines so that medical professionals can determine in the future whether the job has impacted the physical condition of the employee (e.g., hearing impairment). Information gathered as part of the preemployment physical will be used in the final preemployment physical determination, but may also be used in future workers' compensation claims. Therefore, it is important that the medical provider have specific procedures related to preemployment physicals and that medical personnel follow and comply with the procedures.

The laws associated with preemployment physicals are continuously changing; therefore, it is recommended some type of continuous training be considered for personnel directly involved in performing the physicals. When executing a preemployment physical program the following topics should be considered:

- Ensure the job has been offered prior to requiring a physical to be performed.
- The medical history should not be biased toward an understanding of genetic diseases. It is recommended a medical history be tailored only to information that is directly relevant to the job.
- The preemployment physical must be required of all employees, and not those in certain job disciplines.
- Determine whether there is a specified period of time that the employee must be notified of not meeting medical requirements and the offer can be withdrawn (varies by state).

- Be aware of the Americans with Disabilities Act (ADA) requirements and how they pertain to preemployment physicals. The employer is required to make a "reasonable accommodation" in the workplace if the employee qualifies under ADA.

5.3.2 Medical Surveillance

The medical surveillance program is an important part of the routine evaluation of an employee's health. The purpose of a medical surveillance program is to evaluate and monitor the worker's health and determine whether there have been any physiological changes that have occurred since his or her last medical surveillance or employment medical baseline. In addition, information from the medical surveillance program is used to evaluate effectiveness of hazard controls in the workplace. The information may also be used in modifying current chemical, radiological, and biological personal workplace monitoring strategies.

A medical surveillance program should be tailored around the type of work being performed, hazards associated with the work, and environmental conditions that exist when performing the work. For example, if the worker is primarily exposed to chemical hazards, then the medical surveillance program should be tailored to evaluate health/biological parameters (i.e., blood work, urinalysis, etc.) that indicate whether there is any physical impairment occurring as a result of working with the chemical(s). Most chemicals and radionuclides target specific organs for deposition; therefore, the medical surveillance program should be focused on determining whether any harm has occurred to the function of that particular organ. The occupational health provider should be provided a workplace assessment that identifies job duties of that employee and potential exposures to chemicals, radionuclides, or biological or thermal hazards/vectors.

Workplace monitoring data should be provided to and used by the occupational health provider during the routine medical physical. The occupational health provider should possess an understanding of the chemical, radiological, and biological thresholds that are hazardous to the worker and should compare the data against the threshold values in making a final determination as to whether there have been any physiological changes since the last routine medical physical.

The routine medical physical also provides a forum for the worker and the medical provider to discuss any workplace issues or concerns. It is an excellent opportunity for the occupational health provider to reinforce the physical impacts from exposure to chemicals, radionuclides, and biological contaminants and to obtain feedback and identify improvements needed to the medical surveillance program.

The routine medical physical is generally conducted on an annual basis or some other frequency as dictated by company policy following the preemployment physical; however, depending upon the work being conducted, the frequency could be every 6 months or biannually. In addition, depending upon the contaminant of concern, there may be regulatory drivers that specify a periodicity for when the physical is due. The ES&H manager should be aware of these requirements to ensure medically cleared personnel are performing company work.

5.3.3 Injury and Illness Evaluation

Management is responsible to ensure workers are promptly and appropriately provided medical attention following a workplace accident or when the worker reports that he or she has been injured. When an incident occurs or a verbal injury report is made, it is imperative that the worker be given a thorough evaluation to determine whether the injury or illness was caused by operations in the workplace.

Causation is a legal term and is used in occupational health when determining whether or not the injury or illness was associated with work. When performing an injury and illness evaluation the occupational health provider will be determining whether the injury or illness may have "arisen out of employment" or manifested itself in the "course of employment." Specific areas of interest that the occupational health provider will be interested in during an injury and illness evaluation include the following:

- When did the symptoms of the injury or illness first occur, and can the date and time of a specific work activity be tied to the onset of symptoms?
- What are the specific symptoms that the worker is experiencing?
- Have these symptoms been previously experienced, and is there a prior history of injury or illness?
- How did the injury occur?

These questions are tailored toward an occupational safety injury, but a similar set of questions should be asked if there is a potential overexposure to chemicals, radiological, and biological agents. All of this information will be used by the occupational health provider in determining how to treat the worker. This information is also critical to the ES&H manager in making OSHA or other regulatory reporting determinations.

As part of the injury and illness evaluation the event scene should be maintained in case questions arise as part of the medical evaluation. The company can then go back and further investigate workplace and environmental conditions to provide feedback to the occupational health provider. Taking pictures of the scene is very helpful, as well as video surveillance, if available, which is useful in helping to understand how the worker may have been injured. In addition, personal and witness statements and personal or area monitoring data may be useful in understanding environmental conditions. The event scene information can also be used in the reporting determination.

One of the most critical aspects of the injury and illness evaluation process is to ensure information gathered as part of the accident/injury investigation is documented. The time at which the injury was reported is helpful to the occupational health provider in determining causation. The reporting of the injury or illness, along with any other information obtained from photographs, witnesses, environmental data, or workplace observations, is critical to the medical file and ultimately the determination of causation.

5.3.4 Fitness for Duty

Fitness for duty refers to the ability of the employee to be physically able to perform a job or work function. A fitness-for-duty program includes not only the medical physical, but also policies and procedures that provide guidance on how to manage information generated from the fitness-for-duty program. A fitness-for-duty medical exam will determine whether an employee's medical (both mental and physical) condition can reasonably impact his or her ability to perform his or her job. Fitness-for-duty medical exams are generally requested under one of two circumstances:

- The employee is returning to work after being away due to a personal medical condition (e.g., back injury related to a car accident, neurological disease, heart condition).
- The employee is exhibiting signs and symptoms in the workplace that cause supervision to question whether he or she is medically capable of performing his or her job (e.g., drug or alcohol use, potential workplace violence situation, reasonable cause the employee cannot physically perform a task).

When a fitness-for-duty exam is requested the ES&H manager may be directly involved in the process—depending upon the size of the company and organizational structure. Also, if a person is returning back to work from a non-work-related injury or illness, the company is required to reinstate the employee if his or her personal physician has released him or her back to work; however, upon reinstatement, a fitness-for-duty exam can be requested if warranted based on concerns of the company (the concerns must be based on a legitimate issue).

There are a number of legal issues that can arise as part of the fitness-for-duty program. The ES&H manager should work with human resources and the legal department when making employment and workplace decisions that use information generated from the fitness-for-duty program. In particular, there are limitations associated with the medical exam if there is the potential that there could be issues related to the Americans with Disabilities Act.

Fitness-for-duty medical examinations may reveal information about an employee's disability that the employer may not be legally allowed to have access to. If an employee has a disability, an employer can only require a fitness-for-duty medical exam if the exam addresses skills needed to perform the job. Fitness-for-duty programs are also required in specialized industries. For example, the Nuclear Regulatory Commission (NRC) requires nuclear facilities to have fitness-for-duty programs to provide reasonable assurance that nuclear facility personnel are reliable. In addition, other government agencies, such as the Department of Defense and Department of Energy, have reliability and surety programs and standards that must be followed to assure personnel are ready and capable to perform work in hazardous environments.

When making employment or workplace decisions related to information from the fitness-for-duty program it is important to document the process followed, decisions made, and how the decisions are consistent with current company policy. The records generated contain sensitive information and should be managed as such.

5.3.5 Return-to-Work Process

The return-to-work process is used to evaluate a worker after he or she has been away from the workplace. Typical return-to-work programs are used when

- The worker is returning to work after being away from work based on a non-work-related injury or illness.
- The worker is returning to work after being away from work based on a work-related injury or illness.

As with the fitness-for-duty program, the company should have a procedure that defines the process to be followed for return to work of an injured employee (work related or non-work related).

The return-to-work process is initiated by the employee when his or her primary care physician has determined that he or she is able to perform some or all of his or her work duties. The employee then notifies the company he or she would like to return to work on a designated day. When the employee reports for work, he or she should be required to have a medical document that is signed by his or her primary care physician that states what he or she is physically able to perform at work. The occupational health provider will then evaluate the documentation and confirm either full release or partial release (with restrictions) to work.

At the time of the reinstatement it is important that any work restrictions be identified and communicated to the employee's supervisor. It is encouraged that the ES&H manager work with the employee's supervisor and fully understand what work restrictions exist and how, through the normal course of his or her job duties, the supervisor can ensure minimizing the potential to reinjure.

In many cases, because of work restrictions, the employee may only be able to perform part of his or her job duties; however, the supervisor and the ES&H manager should work with the employee and identify alternative work that can be as productive and rewarding. In particular, if the work restriction is related to a work-related injury, both the company and the employee will benefit from having the employee return to work early—as long as the employee's work is monitored to minimize the potential for reinjury. In addition, the ES&H manager and supervisor should work together to acclimate the worker back to working a full day. The supervisor should periodically monitor work activities to ensure the employee is not performing work that has not been authorized given his or her physical limitations.

All information generated as part of the return-to-work process should be documented and kept as part of the employee medical file to ensure appropriate actions were taken and documented, and the file readily retrievable in case of future litigation.

5.4 Prevention and Wellness Programs

Prevention and wellness programs offered by employers are designed to promote health and prevent disease. Prevention and wellness programs assess participants' health risks and deliver tailored educational and lifestyle management interventions designed to lower risks and improve outcomes. These programs can be essential in

- Reducing health care costs and claims
- Improving health and wellness of the employee
- Encouraging and promoting new membership in the program along with satisfaction with program results

Prevention and wellness programs generally are a combination of educational and organizational activities that target promoting a healthy lifestyle both at work and away from work. These programs typically consist of health education, screening, and interventions designed to improve workers' behavior in order to achieve better health. Examples of some prevention and wellness programs include the following:

- Weight control programs
- Smoking cessation classes
- Stress management classes
- Flu clinics
- Information on health topics, such as heart disease and foods that promote a healthy lifestyle
- Reduced gym membership pricing
- Employee recognition programs

Certain types of wellness programs offered through employment-based group health plan coverage may be required to meet standards under the Affordable Care Act. In most cases, the ES&H manager is involved in the implementation of such programs. Often the ES&H manager works with the occupational health provider to develop safety and health campaigns that will target various aspects of the company's prevention and wellness programs. Almost all occupational health providers offer these programs as part of the medical contract, and the ES&H manager is encouraged to work with the provider to reduce the number of future medical claims.

5.5 Injury and Illness Case Management and Reporting

Injury and illness case management and injury reporting are interrelated in the execution of an occupational health program. There are several different facets of case management and injury and illness reporting that the ES&H manager will be responsible for ensuring is appropriately executed, including the following:

- Employee management of the injury and illness from initiation to closure from a medical perspective
- Workers' compensation
- Injury and illness classification for OSHA reporting purposes

All of these topics are related to management of employees who are injured in the workplace and covered under OSHA. Reportability requirements are different for radiological incidents and uptakes and should be followed as defined in the regulations for an occupational program for radiation.

From the time the employee first reports the injury the employee supervisor or manager should be engaged and accompany the injured employee to the occupational health provider. It is the company's responsibility to ensure all information is communicated to the occupational health provider so that the right medical treatment can be provided. Because there is generally a company representative that accompanies an injured employee when seeking medical attention, it is important that the company representative understand his or her limitations when it comes to obtaining employee medical information because of the Health Insurance Portability and Accountability Act (HIPAA).

HIPAA is a federal law that protects individually identifiable health information that relates to

- The individual's past, present, or future physical or mental health or condition
- The provision of health care to the individual
- The past, present, or future payment for the provision of health care to the individual, that identifies the individual, or for which there is a reasonable basis to believe it can be used to identify the individual

Individually identifiable health information includes many common identifiers (e.g., name, address, birth date, social security number). Information that is gathered as part of the medical treatment process is specifically covered under HIPAA. The employer is, however, allowed to have access to enough information to understand the type of treatment provided, work restrictions, and any additional information the employer would need to safely manage the employee with respect to job duties.

The ES&H manager should ensure statements from witnesses and the employee himself or herself (along with scene investigation information) are obtained for use in the case management and injury reporting process. If the occupational health provider determines that injury was caused by the work being performed, then the employee is entitled to file a workers' compensation claim.

The primary intent of the workers' compensation program is to ensure medical expenses and some form of salary are provided to workers who are injured on the job. Companies are either self-insured or pay for insurance that covers their liability costs should an employee become injured and it was caused on the job. Once an employee files a workers' compensation claim, the insuring party will make a determination as to whether it deems the injury to be compensable.

Different insuring parties use different criteria in their determination as to whether a claim is compensable, and there is variability between states as to what may qualify as compensable. Should the claim be denied compensability by the insuring party, the injured employee has the right to challenge the determination. Most states have compensation boards that review claims and may reverse the compensability determination and require costs to be reimbursed to the employee.

The compensability determination pertains to whether the employee injury will be covered within workers' compensation criteria (varies by state). The reportability determination pertains to whether the injury or illness meets the injury reporting criteria within OSHA. Often an injury may meet the threshold to be compensable but not reportable. There are also some cases where an injury may be reportable (i.e., loss of consciousness due to blood draw), but would result in no claim being filed for compensability. This is an important distinction because many people believe they are the same and inadvertently report injuries to a regulatory authority that are not required to be reported.

Reporting of injuries under OSHA is fairly well defined. Several criteria must be met when determining whether an injury is reportable on the OSHA 300 log and in some cases to OSHA itself:

- Was the injury or illness causation determination related to work?
- Is the case new?
- Does the injury or illness meet the general recording criteria of
    - Death
    - Days away from work
    - Restricted work or transfer to another job
    - Medical treatment beyond first aid (as defined under 29 CFR 1904.7)
    - Loss of consciousness

For both compensable and reporting purposes there are generally specific timeframes within which a reportability determination must be made (i.e., within 7 days of the determination of causation for injury reporting and 14 days for compensation determinations from the time the claim was submitted); however, the timeframes for determining compensation will vary depending upon the state.

5.6 Challenges of Managing an Occupational Health Program

There are many challenges that may be encountered when managing an occupational health program. Occupational health providers are highly trained personnel in various specialized fields. Oftentimes their training and experience is in the hospital or private medical practice setting, not general industry. Their focus is on patient care and ensuring standards and ethics are upheld in caring for the patient. Consequently, there is a balance that medical providers must consider when contracting their services in supporting commercial businesses and government. Although patient care continues to be their primary focus, they must also consider how to interact with and look out for the liability interests of the company.

The ES&H manager can work with the medical provider to better understand work operations and company policies and procedures that will effectively integrate both the interest of the patient and the company. The ES&H manager may be challenged with reinforcing roles and responsibilities associated with work performed by the occupational health provider. It is important that the medical staff understand their skill set is limited to treatment aspects of occupational health programs, but they are not trained in other aspects of the company business and must learn how to recognize when an issue is outside their area of expertise and training.

The ES&H manager is generally responsible to ensure the occupational health provider is conducting business in accordance with regulations and standards. As with all contracted or specialized services, there should be frequent oversight of the workplace. The ES&H manager should ensure all required medication and clinic inspections are being completed, documentation is accurate and up-to-date, and that there is appropriate management of personnel and documentation. Routine walk-throughs and inspections are also recommended to ensure appropriate medical care is provided to workers.

5.7 Summary

Safety and health professionals (including those related to other disciplines of ES&H) manage worker risks from the immediate work environment, while the health care professionals assess and medically treat the worker, whether it be for injury or illness prevention or for injury or illness beyond first aid. When managing an occupational health organization, the ES&H manager must be aware there are several facets of the program that will require his or her attention and management skills.

The size and organizational structure of the occupational health group will vary depending upon the type of operations being conducted (level of health risk) and the size of the company. It is important for the ES&H manager to work with the physician in communicating roles and responsibilities and expectations of the medical staff. There are a number of functions the medical staff performs as part of an occupational health program, such as preemployment physicals, medical surveillance program evaluations, injury and illness evaluations, fitness-for-duty and return-to-work evaluations, prevention and wellness programs, and injury and illness case management.

There are many challenges the ES&H manager may encounter when managing an occupational health program. However, many of these challenges can be overcome by working with the occupational health provider, the risk manager, and the supervisor to ensure that each employee is provided optimal service and the risk to the company is minimized.

6

Environmental Protection

6.1 Introduction

The discipline of environmental protection (EP) serves as a conduit for companies to facilitate compliance to the regulations established by the Environmental Protection Agency (EPA) and other agencies that have enacted regulations designed to preserve the environment. The EPA was established to protect human health, reduce and control pollution to the environment, and ensure that toxic substances are properly disposed. The EP functional area is critical to the goal of protecting and sustaining the world's natural resources.

Environmental safety and health (ES&H) managers and professionals working in this field must be prepared to deal with the many regulations enacted and enforced by the various federal, state, and local agencies. The involvement of the regulatory agencies coupled with the potential fines that can be imposed seems to provide the best incentives for companies to comply with the requirements. The penalties that can be imposed against an individual or a company can range from monetary fines up to imprisonment when noncompliances are discovered by the regulators. Establishing and implementing an environment protection program is an important management responsibility that can yield great rewards for the environment and the company.

6.2 Organization Structure and Design

The environmental functional area (EFA) is an integral part of an ES&H program. There are various ways to construct or design the EFA organization in a way to meet the goals of the company.

When designing the organization, first determine which of the environmental regulatory requirements are applicable to the business along with the enforcement structure. For example, which agency is responsible for enforcement: federal, state, or local government, or a combination of the three? An effective program will have a strategy and a process in place to perform the following at a minimum:

- Investigation and examination of workplace hazards that may impact the environment
- Training and education of the workforce concerning job-related risks to the environment
- Recommendation of improvement measures to support a safe work culture
- Measures to protect the environment
- Continuous improvements programs
- Metrics to gauge and demonstrate compliance to environmental regulations and permits
- A public outreach program design to educate the public on measures the company has taken to ensure safety of the public
- A well-defined audit program

A typical EFA organization may resemble the structure shown in Figure 6.1. Each of the functions in Figure 6.1 will have components such as data collection and analysis, monitoring, reporting, and continuous improvements. The program/project support staff represent the portion of the organization that is matrixed to the programs or projects and provide direct support in the area of environment protection. This portion of the staff is integrated and can report directly to line management.

The program/project staff oftentimes possess a broad depth of experience and knowledge of environmental protection regulations and procedures, and are tasked with assisting line management in implementing the environmental regulatory requirements for the facilities and projects. These professionals need to maintain direct contact and receive support from the functional area to ensure consistency of program implementation.

The functional subject matter experts (SMEs), the portion of the organization that is not matrixed to line management, are responsible for

- Serving as the regulatory knowledge center, keeping abreast of new regulations and regulatory changes
- Communicating with the regulators on behalf of the company
- Preparing and submitting permit applications
- Revising and tracking permit applications
- Monitoring, analyzing samples, and reporting results to management and the regulators

The environmental monitoring team provides direct support to the functional area team to ensure that samples collected can be used to demonstrate

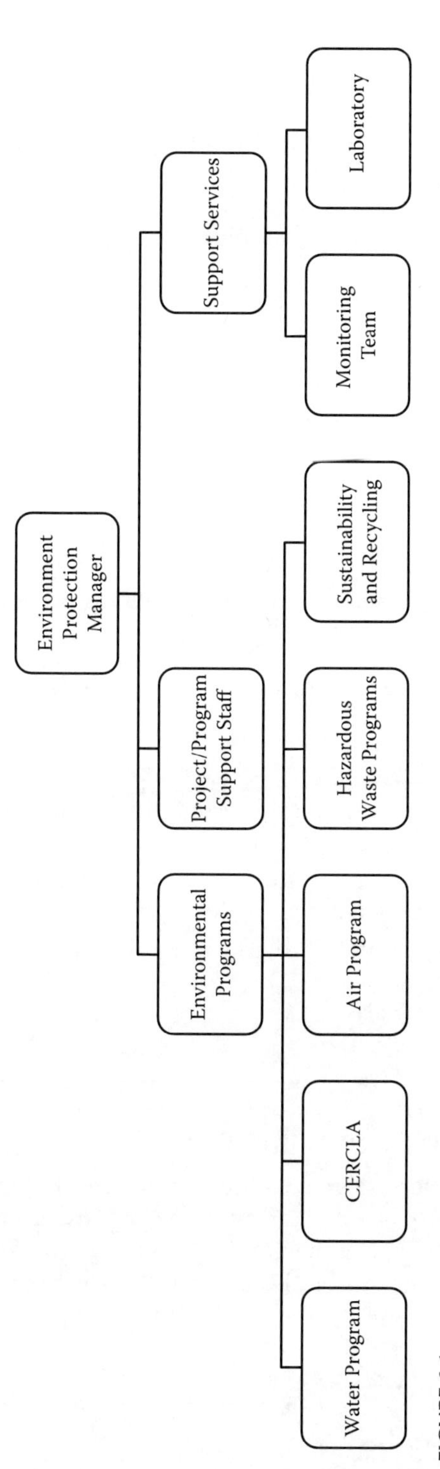

FIGURE 6.1
EFA structure.

compliance and meet permit requirements. A sampling strategy and schedule are recommended to be developed to ensure that samples are collected and analyzed using prescribed methods. In addition, a schedule is useful to ensure that reports are prepared and transmitted to the regulatory agency in a timely manner.

It is necessary and required by regulation to have a credible and oftentimes certified laboratory to analyze the samples collected. Laboratory services can be provided internally or by an off-site laboratory. If analysis is performed internally, it is necessary to ensure the following are addressed:

- Ensure that the staff is trained and skilled to execute laboratory procedures
- Implement a quality control program
- Determine if the laboratory can meet permit limits for contaminants of concern
- Determine if the internal lab can handle the sample volume and turnaround time needed
- Determine if a contract with an internal lab is necessary to handle sample overload

If the decision is made to contract with an external laboratory, the following should be considered:

- That the laboratory has a quality control program
- That the laboratory can meet permit limits for contaminants of concern
- Capability of the laboratory to analyze samples and provide results in a timely manner as defined by the company
- Credibility of the laboratory by the regulatory community and outside stakeholders

6.3 Regulatory Structure and Drivers

As previously stated, there are many regulatory drivers targeting protection of the environment. A primary regulatory driver for environmental regulatory compliance is from the Environmental Protection Agency (EPA), which began on December 2, 1970. Responsibilities of the EPA include (but are not limited to) the following:

- Conducting research.
- Developing sound standards to protect and enhance the nation's environmental posture.

- Enforcement of environmental standards.
- Administration of the Environmental Response, Compensation and Liability Act (Superfund). This act is designed to aid in the restoration of toxic waste sites by enforcing cleanup.

Other regulatory drivers designed for the protection of the environment include regulations enacted by the Fish and Wildlife Services (FWS), the U.S. Army Corp of Engineers (USACE), the National Marine Fisheries Service (NMFS), and state and local agencies.

Although the environmental regulations are developed largely by and enforced by the EPA and other federal agencies, there are many other local and state entities that have been given authority by these agencies to also enforce these regulations. In addition, state and local entities can add additional requirements to the standard flowed down to them to enforce by the EPA and other federal agencies.

What does this mean to a project or company and the ES&H manager? This means that there can be a marked increase in regulatory oversight depending on where business is being conducted, the type of business, and the established compliance history of the company, thereby increasing the interaction by company personnel with compliance officers. Since there are several regulatory drivers that can be applicable to the business, before developing environmental programs and conducting business the ES&H manager should ensure that all regulations (federal, state, and local) are consulted to verify and validate that business will be compliantly conducted.

Recognizing that there are many environmental regulations that can impact the way business is conducted, knowledge of these regulations is required to ensure compliance. A segment of those regulations is referenced in Table 6.1. The regulatory drivers discussed in the table are not all-inclusive. These regulations are highlighted to serve as a means to familiarize the manager on some form of or the types of regulations that may need to be considered when developing business practices and while conducting daily operations.

6.4 Waste Management

Handling the waste that is generated as a result of conducting business can be challenging and expensive depending on the type of waste being generated. There are specific requirements for managing all types of waste, such as paper products versus hazardous waste. A solid waste management program is necessary to ensure waste is properly disposed and to avoid the

TABLE 6.1

Environmental Regulations

Regulation	Purpose
Comprehensive Environmental Response Compensation and Liability Act (CERCLA) 1980	The law imposed a tax on the chemical and petroleum industries and provided broad federal authority to respond directly to releases or threatened releases of hazardous substances that may endanger public health or the environment.
Clean Air Act (CAA) 1970	Designed to regulate air emissions from stationary and mobile sources. The law authorizes EPA to establish National Ambient Air Quality Standards (NAAQS) to protect the public and public welfare and to regulate emissions of hazardous air pollutants.
Clean Water Act (CWA) 1972	Establishes the process for regulating the quality standards for surface water and pollutants discharge into the waters of the United States. The CWA made it unlawful to discharge any pollutant from a point source into navigable waters, unless the source is permitted.
Emergency Planning and Community Right-to-Know Act (EPCRA) 1986	Established to help communities plan for and deal with emergencies involving hazardous substances. EPCRA requires industries to report to federal, state, and local governments the storage, use, and release of hazardous chemicals.
Endangered Species Act (ESA) 1973	Passed by Congress with the expressed purpose of providing protection and conservation of threatened and endangered plants, animals, and their habitats to avoid extinction. Endangered species can include insects, birds, flowers, grass, reptiles, mammals, and trees. The ESA is administered by the Fish and Wildlife Services (FWS) and the National Marine Fisheries Service (NMFS) with split responsibilities.
National Historic Preservation Act (NHPA) 1966	The act is intended to preserve historical and archaeological sites within the United States. A registry of historic places and landmarks was generated as a result of the act.
National Environmental Policy Act (NEPA) 1970	NEPA places requirements on federal agencies to integrate environmental aspects into their business decision-making processes. It forces companies to consider environmental impacts of their business proposals and alternatives to those decisions when needed.
Resource Conservation and Recovery Act (RCRA) 1976	RCRA applies to the management of solid waste such as garbage, hazardous waste, and underground storage tanks used to store petroleum and chemical products. The goal of RCRA is to protect human health and the environment from the potential hazards resulting from the disposal of waste, conserve energy and natural resources, reduce the amount of waste generated, and ensure that wastes are managed appropriately.
Toxic Substance Control Act (TSCA)	TSCA creates the regulatory directives to collect data on chemicals in order to evaluate, mitigate, and control the potential risk during manufacture, processing, and use. Through TSCA, EPA has mandated that companies provide chemical inventory.

costly penalties that can be assessed by the regulators. Waste management techniques include (but are not limited to) the following:

- Recycling and reuse of material where feasible
- Substitution of environmentally friendly products for those that have shown negative impact on the environment
- Just-in-time purchasing or inventory control—purchase only the material that is needed to avoid the potential of the need to discard material that has expired

6.5 Environmental Permits

Environmental permits serve as authorization for implementation of a specific environmental regulation. These permits address management of issues related to company processes or systems that may generate harmful emissions or discharges. A large percentage of the environmental regulatory requirements that will be applicable to a facility will be documented in the form of an environmental permit. These permits are also referred to as operating permits that the regulatory authorities issue to emission sources that will allow the source to legally operate in accordance with regulatory requirements.

Permits are designed to clarify what facilities must do to control pollution and protect the environment and to improve compliance. These permits:

- Are legally enforceable documents
- Contain specific operating parameters that must be followed
- May be issued by federal, state, or local entities

The permitting process can be simple or complex depending on the type of permit, the federal, state, and local requirements, and the process or system that is being permitted. Therefore, once it has been determined that a permit is needed, the path forward to secure the permit should begin. The process can take from weeks to years depending on the permit request, the interest of the public, and the regulatory agency's ability to issue the permit. A simplistic diagram depicting the permitting process is shown in Figure 6.2. The process can be different depending on the attributes listed above.

6.6 Regulatory Compliance and Reporting

We have established that environmental regulations are written to protect the environment through protection of the air we breathe, the water we

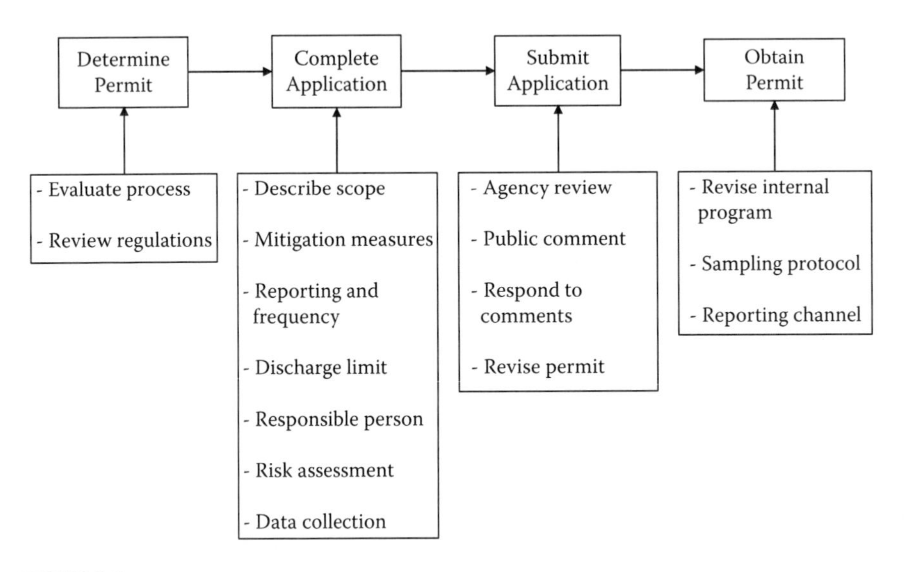

FIGURE 6.2
Permitting process.

drink, and the soil that provides the source of nutrients for food. Each permit obtained by a company has specific requirements that must be followed as written. It is important to establish program and project structures and systems to ensure that the requirements of these permits are followed and that compliance can be demonstrated when requested. Penalties for noncompliance can include the following:

- Monetary fines
- Compliance-driven and regulatory enforced changes to company processes and systems
- Increased regulatory visits
- Imprisonment
- Poor reputation in industry, which can lead to a decreased ability to secure future work

There are other penalties that a company can pay that can impact the bottom line. These penalties are not as visible and may occur over time. These penalties can include the loss of trust from employees, stakeholders, customers, and shareholders. Loss of trust by the customer can impact business and thereby impact the financial viability of the company.

6.7 Environmental Sustainability

The concept of environmental sustainability involves making decisions and taking actions that are in the best interests of protecting the capability of the environment. Simply put, environmental sustainability is not just about reducing the amount of waste produced or the amount of energy used; it is about making responsible decisions that will reduce negative impacts of businesses on the environment.

When conducting business it is necessary for the ES&H manager to consider the impact that the business will have on the environment and to look beyond making short-term gain decisions that may negatively impact the environment and look at the long-term impact. The management team should demonstrate their support by being responsible in their business decisions and adopt a life cycle approach to sustainability by considering the impact of the projects being designed or products being developed and the long-term impact they may have on the environment.

To ensure that sustainability of the environment is always on the minds of the leadership team and the workers, it is a good idea to set defined sustainability goals and work toward achieving them. Example sustainability goals may include the following:

- Reduce the amount of water used by 10%.
- Reduce the amount of carbon by-products emitted by 15%.
- Increase the volume of recycled material by 5%.

6.8 Employee Involvement

A company's environmental management program will not be effective without buy-in and involvement by management and the employees. Management sets the tone for program acceptance, and generally the worker will follow. This means that senior management and the ES&H manager should not only verbalize their support for protecting the environment, but also demonstrate their support through their actions.

Employees will follow a credible and trustworthy management team that is charting the course through their demonstrated support. Management is responsible for program development, implementation, and compliance. An environment management program will not be effective without employee engagement. Employees must be trained or briefed on the environmental management strategy and the expectation for compliance. Many companies have employee-led environmental committees that contribute to not only goals and objectives of the program, but also environmental sustainability initiatives.

6.9 Environmental Interest Groups

Environmental interest groups continue to gain popularly across the country with the intended mission of protecting the environment from adverse impact of business operations. These groups are at times referred to as environmental "watch dog" groups. Group members monitor activities performed by companies that may have an impact on the environment, monitor applications for permits, to include renewal and new permits requested, and monitor company performance through information published by the media and on the company external websites. These groups can impact the company in various ways, such as how quickly a permit is issued by the regulators and the perception by the public on the environmental responsibility of the company.

6.10 Environmental Audits and Inspections

Conducting audits and inspections to evaluate the company's environmental compliance posture not only is a good business practice, but also, in some cases, is required by the regulations. The information gained from conducting these inspections can be used to continuously improve and modify the company program as needed. The purpose and benefits of conducting environmental audits and inspections include the following:

- Demonstrating compliance to the regulations
- Determining liabilities
- Tracking and reporting of compliance cost
- Increasing environmental awareness
- Tracking accountability
- Improving compliance
- Decreasing fines and external audits

The knowledge gained from inspections should be used to continuously improve and, in some cases, redefine the environmental program when necessary. It is helpful to have predeveloped checklists that can be used by the worker to conduct some routine inspections. A trained auditor or audit team should be used when conducting a review of the flow-down and compliance to requirements.

There are two primary types of audit that are used in evaluating the status and effectiveness of an environmental management program: environmental management system and compliance audits. Both of these approaches are discussed in brief below.

6.10.1 Environmental Management Systems

An environmental management system (EMS) is the protocol used to manage the environmental protection systems established to ensure protection of the environment and compliance with applicable regulatory requirements. These systems are strategically planned, comprehensive, documented, and systematically aligned.

There are specific protocols that should be followed when auditing the entire environment management system. The type of audit that is expected to be performed, and is most beneficial in this situation, is a complete audit of the entire environmental system. Utilizing the International Organization for Standardization (ISO) 14001 standard can be helpful in providing guidance on the appropriate procedure for performing EMS audits and qualification of the audit team. The ISO is an international standard designed to ensure products and services provided by a company are safe and of good quality.

The ISO 14001 standard specifies requirements for an environmental management system to enable an organization to develop and implement a policy and objectives that take into account the legal requirements and other requirements that a company must meet, and the information about significant environmental aspects. It applies to the environmental aspects that the company identifies as those which it can control and those which it can influence. It does not itself state specific environmental criteria. The ISO 14001 process can be used even if the company is not ISO 14001 certified or has no desire to seek the ISO certification.

6.10.2 Compliance Auditing

Ensuring compliance is not an easy task in today's environment due to the amount and complexity of some of the regulations. Compliance auditing is another auditing technique used to check actual performance against the regulatory requirements. This method is a great tool for the ES&H manager to determine whether a process is in compliance with specified requirements. Compliance auditing is also a great tool to use to prepare for an audit of the environment management system.

Generally, when conducting these types of audit, it is common to closely assess one or more elements of the program using a checklist as opposed to the entire program in one assessment. An example of a checklist that can be used for inspecting a hazardous waste storage area is shown in Table 6.2.

TABLE 6.2

Hazardous Waste Storage Area Checklist

Reviewer: ______________________ Date: ____________ Time: ________

Location: ______________________

Satellite accumulation area (<55 gallons of hazardous waste or <1 quart of acute hazardous waste)

Activity	Yes	No	Comments
Housekeeping maintained			
Waste accumulated at or near the point of generation			
The container in good condition with no visible leakage			
The container labeled "Hazardous Waste" and generator information			
The waste stored in a container that is compatible with the waste			
The container kept closed			

90-Day Storage Area

Activity	Yes	No	Comments
Housekeeping maintained			
Container marked with the date accumulation started			
Container labeled with "Hazardous Waste" and generator information			
Container is in good condition and no visible leaks			
The waste stored in a container that is compatible with the waste			
Container kept closed			
Incompatible materials in a separate area (50 feet away)			
Weekly inspection performed			

Other checklists can be developed and used to add ease and consistency to the compliance auditing process.

6.11 Summary

In this chapter we have tried to provide insight on the tremendous responsibility the EFA has in developing and maintaining a compliant environmental management system and process. We have introduced concepts that are commonly encountered and attributes that one may not intuitively consider when conducting business. Key points that senior management and the ES&H manager must remember in order to increase their compliance posture and reduce the risk of being subjected to fines and penalties and to demonstrate responsibility in protecting the environment include the following:

- Ensure that a sound environmental management program is in place
- Ensure that the environment staff is knowledgeable in the environmental regulations and what regulations are applicable to the business
- Conduct business responsibility in a way that the environment is not negatively impacted
- Demonstrate support for environmental policies and procedures

The regulatory requirements that were mentioned by no means represent all of the regulatory requirements that must be followed. The point is that it is important for the ES&H manager to know the regulatory requirements that the business will be subjected to in order to maintain a compliant posture.

7

ES&H Program Support

7.1 Introduction

The ES&H organization plays a vital role in the success of the company in achieving its goal and mission. Whether the company is tasked with building automobiles, constructing a factory or home, operating a manufacturing or chemical business, or conducting research and development, the primary mission of the company cannot be achieved without each organization performing its specific functions, including ES&H. Within the company ES&H not only supports the day-to-day operations, but also is heavily involved with the following:

- Cost-saving initiatives
- Reducing workers' compensation costs and potential legal liability
- Assisting the company in being competitive in its safety and environmental statistics
- Improving the overall working environment for workers

Just as an ES&H organization is the backbone to performing work in a safe and compliant manner, the program support group of the ES&H organization serves as the cornerstone to program development and field implementation.

The program support group is an integral part of the ES&H organization that is chartered to support the entire organization in developing and managing the support functions that are necessary to enable regulatory compliance while supporting the people aspects of the organization and operations. The program support group acts as the hub for the ES&H program and field implementation, and if the program support group is successful in meeting commitments, they will enable efficient and effective ES&H services across the company.

The program support group typically consists of personnel who frequently interface with outside regulatory agencies and stakeholders; therefore, they can be instrumental in shaping the perception of the company as being proactive and a protector of the workers and the environment. Most workers, supervisors, and managers external to ES&H are not aware of all

FIGURE 7.1
ES&H program support group organizational structure.

the functions performed by the program support group. The program support group manages a wide variety of activities, including:

- Assurance of compliant company operations to ES&H requirements
- Focused communications from the ES&H organization
- Injury and illness reporting
- Business and financial support
- Program management
- Chemical safety

A typical organizational structure for the program support group is pictured in Figure 7.1. Many of these disciplines will be discussed in greater detail in the sections below.

The size of the program support group will vary depending upon the following:

- The size of the company and the ES&H organization
- The regulatory requirements
- The size of the chemical inventory
- The hazard profile of the company
- The type of work performed

The smaller the company, the more functions one person may perform, whereas in a large company, due to the level of responsibilities, there may be more specialized subject matter experts or more aspects of ES&H.

7.2 Performance Assurance Office

The performance assurance office consists of experts with specialized skills to review and assess compliance to company policies, procedures, and regulations. The performance assurance office also participates in and ensures the proper reporting of issues or incidents to the applicable regulatory agencies and the customer. The work that the performance assurance office performs is related to regulatory compliance or continuous improvement. Below is a listing of several work activities that involve the performance assurance office; remember that the performance assurance office for the ES&H function is limited to only the ES&H organization:

- Performance of required assessments, audits, and surveillances
- Identification and evaluation of regulatory noncompliance screenings if required
- Evaluation and resolution of issues and associated corrective actions
- Trending of metrics and indicators

The performance assurance office evaluates from an independent perspective work that is performed to ensure it is being conducted in accordance with procedures and requirements. Issues that are identified and required to be corrected by the ES&H organization are evaluated by the performance assurance lead, and resolution is monitored to ensure that the solutions are protective and meet the intent of the corrective action. It is the performance assurance office that generates metrics and indicator information for ES&H that is then routinely evaluated by management to determine whether a positive or negative trend exists. The performance assurance office for ES&H may also perform noncompliance screenings (generated from company-wide and ES&H organizational issues) to identify whether there may be a regulatory noncompliance that is required to be reported to a regulatory agency. This office has responsibility for overall management of all tasks within its purview and frequently interacts with field and other programmatic ES&H personnel.

7.3 Communication Office

The communication subject matter expert (SME) or group is responsible for developing and administering all communications related to the ES&H organization and distributing information to improve health and safety of the workforce. Example communications that are disseminated include (but are not limited to) the following:

- Safety and health, radiation protection, and environmental newsletters
- Statistics related to implementation of all programs
- Lessons learned

- Changes in policies and procedures
- Regulatory changes
- Assistance in the development of external communications in conjunction with the company's public affairs office

Dissemination of information is a routine activity of any ES&H organization. It is an expectation that the ES&H organization communicate information of relevance and importance to the worker, as well as to management, on topics related to safety and the environment, and it is the responsibility of the communications SME to ensure those communications occur. Included in this work activity is the dissemination of lessons learned, which will assist in learning from past events to prevent reoccurrence of similar events in the future and in recognizing unsafe behaviors or acts.

The communications SME is a person specifically trained in the development of various types of written documents to communicate and can appropriately assist in disseminating many types of information, including charts, graphics, statistics, and written text (i.e., newsletters). In most cases, the communications SME has the training and skills to write and present information in such a manner that it is easily understood by people at all levels and is favorably viewed and leaves a positive impression in the minds of the reader of the company.

The communications lead acts as an extension of the company's public affairs office and interfaces with it on a routine basis. The function of the communications SME can be of great importance when the company is large or is part of a larger corporation. They can be extremely beneficial when a regulatory enforcement action occurs or when an accident occurs and will work with the company's public affairs office on the appropriate message to be disseminated. Frequently the information generated by the ES&H organization is sensitive (could be subject to future legal litigation), so it is important to have a centralized person through which all formal communications are managed.

7.4 Injury and Illness Management and Reporting Office

The injury and illness management and reporting SME is tasked with the responsibility of ensuring injury case management is handled in accordance with the Occupational Safety and Health Administration (OSHA) standards and company policies, as well as may manage the workers' compensation program for the company. This person works closely with the occupational health provider and any third-party administrator that may be contracted to support the company's workers' compensation program. Personnel performing this function generally have a good understanding of the OSHA reporting requirements and workers' compensation laws.

When an employee becomes injured on the job, the injury and illness reporting SME is contacted and assists throughout the worker's injury and recovery process. Listed below are some of his or her responsibilities:

- Work with the employee in filing the appropriate workers' compensation form
- Obtain the medical information from the occupational health provider on the extent of the injury and treatment
- Completion of all OSHA documentation once a determination is made by the ES&H manager of the SME that the injury is work related
- Routinely monitor and evaluate how the injured employee is recovering
- Assist in the return-to-work process
- Interface directly with the third-party administrator for the workers' compensation program and respond to informational requests
- Work with management in keeping them abreast of open work-related injury cases and overall cost to the company from work-related injuries

The injury and illness reporting office is oftentimes the person who most frequently communicates with an injured employee and at times the medical provider. He or she tends to become the face of the company should an employee be off work for an extended period of time. Because of that relationship, it is recommended that the ES&H manager routinely meet with the injury and illness SME to stay informed of how recovery of work-related injuries is progressing and to understand whether legal counsel may need to be consulted in case of future litigation claims.

It is a good practice for the ES&H manager and the injury and illness reporting lead to meet with senior management of the company so that the company leadership can understand costs associated with worker injury, recovery time, and possible retraining of injured workers who are no longer able to perform their previous job. Should the workers' compensation costs become significant, senior management of the company, as well as the corporation, need to address causes of the cost increases and adjust their management approach accordingly.

7.5 Business and Finance Office

As with other types of organizations, the ES&H organization is expected to conduct business within the guidelines of the company and within its allocated budget. The ES&H business and finance office assists the organization by carrying out functions such as

- Maintaining the organizational budget
- Budget planning and forecasting
- Procurement of supplies and equipment
- Maintaining an organizational schedule directly tied to programmatic initiatives and milestones
- Assisting with procurement contracts (e.g., external lab services, repair of equipment)

The scope of work that the business and finance lead performs may not seem very extensive; however, the functions he or she performs are extremely vital to the livelihood of the ES&H organization. Appropriate and effective management of the organization's budget is critical to ensure that work currently scheduled can be funded and accomplished. In addition, future planned work that is scheduled to be conducted later in the year cannot be performed if funding is not available.

The focus of the company is to make money and generate shareholder profits; any sort of mismanagement of funding is generally not tolerated. In some companies profit from the client is directly linked to completed work as scheduled; therefore, an accurate ES&H project schedule, that is resource loaded, can be critical for ensuring that the company can maximize safe completion of work. The ES&H manager should routinely schedule status meetings with the business and finance lead to review scope, cost, and schedule considerations for the organization. In addition, a synopsis of this information should be routinely shared with ES&H personnel so that everyone within the organization understands funding constraints and the planned work scope for the year.

There are a number of items that are routinely procured as part of the ES&H organization. Items such as computers, instrumentation, subcontracted services, and personnel are just a few examples of procurements related to an ES&H organization. The business and finance office assists in assuring the procurement process is progressing in a positive way and that funding is appropriately allocated to ensure a successful procurement and bills are paid on time to ensure accuracy of the organization's budget.

7.6 Program Management Office

The program management function serves as the fundamental foundation by which the ES&H organization operates. There are a number of work activities that may be associated with the program management function, and they can be quite extensive. Listed below are a number of those activities:

- Development and maintenance of policies and procedures
- Development and maintenance of roles, responsibilities, accountability, and authority (R2A2) for the ES&H organization
- Completion of annual reports
- Corporate and client reporting of ES&H information
- Routine interface with regulators
- Management of company-wide campaigns and initiatives
- Responding to requests by management, other departments, regulators, and outside stakeholders

Policies and procedures are a means to help create the framework used to manage organizational practices. Policies and procedures are necessary to:

- Communicate the vision for the company and the organizations
- Communicate acceptable practices and expectations for organizational members
- Guide consistency in program implementation
- Help add a sense of order for daily operations
- Serve as a road map to guide operations without constant communication with management

The operational aspects of an ES&H organization are supported by the ES&H program management function through development and maintenance of policies and procedures that dictate the way the business is organized and operated. These policies and procedures are also maintained to ensure accuracy, consistency, and regulatory compliance. In addition, it is the responsibility of the program support function to ensure that the procedures are kept current with regulatory changes and changes in the compliance posture of the ES&H organization and the company. An important set of documents that is generated and maintained by this group includes the following:

- Policies for environment safety and health
- Procedures related to all aspects of the ES&H organization, including operational procedures, reporting procedures, notification procedures, response to emergencies (to list a few)
- Forms that are routinely used when executing the procedures
- Databases that may be required as part of programmatic activities

Included in the documents that are developed and maintained are the roles and responsibilities for the organization. Tables 7.1 through 7.4 list

TABLE 7.1

Safety and Health Roles and Responsibilities

Responsibility	Role
Ensure programs are implemented in the field that protect workers from industrial and chemical hazards during operations	Safety representative
Communicate and work with management and the workforce in implementing these programs to ensure the company meets established regulatory standards during operational activities	Safety representative Employee safety lead
Advise management as to when working conditions do not meet established standards	Safety representative Employee safety lead
Be an advocate for the company, with the workforce, in promoting safe work practices	Safety representative Employee safety lead Safety and health manager
Ensure safety representatives are performing their job in the field	Safety and health manager
Routinely monitor conditions during operations to ensure the company is meeting regulatory requirements	Safety and health manager
Ensure funding is allocated to adequately perform the safety and health function during operational activities	Safety and health manager

TABLE 7.2

Radiological Protection Roles and Responsibilities

Responsibility	Role
Perform radiological monitoring of radiological areas, in accordance with written survey instructions and regulatory requirements	Radiological control technician
Identify changing or abnormal radiological conditions in the field	Radiological control technician
Monitor radiological conditions when operational activities are being performed and document such conditions	Radiological control technician
Provide health physics support in the planning and execution of radiological operational activities	Radiological protection professional
Evaluate data generated during field monitoring of radiological areas and recommend changes as needed to ensure personnel are being protected and regulatory requirements are met	Radiological protection professional
Routinely monitor conditions during operations to ensure the company is meeting regulatory requirements	Radiological protection manager
Ensure funding is allocated to adequately perform the radiological protection functions (such as instrumentation and monitoring during operational activities)	Radiological protection manager

TABLE 7.3

Environmental Scientist Roles and Responsibilities

Responsibility	Role
Perform environmental sampling of environmentally regulated areas, in accordance with written sampling plans and procedures and per regulatory requirements	Environmental science technician
Identify changing or abnormal environmental conditions in the field	Field environmental lead
Monitor environment conditions when operational activities are being performed and document such conditions	Field environmental lead
Evaluate data generated during field monitoring of environmental parameters and recommend changes as needed to ensure personnel and the environment are protected and regulatory requirements are met	Field environmental lead
Provide environmental protection support in the planning and execution of operational activities	Environmental scientist
Routinely monitor conditions during operations to ensure the company is meeting regulatory requirements	Environmental protection manager
Ensure funding is allocated to effectively perform the environmental protection functions (such as instrumentation and monitoring during operational activities)	Environmental protection manager

TABLE 7.4

Health Services Roles and Responsibilities

Responsibility	Role
Initial medical responder to worker injuries and work accidents involving personnel requiring medical attention during operational activities	Paramedic
Monitor physiological parameters while operations are being conducted	Paramedic
Perform routines as part of medical program in support of operations (i.e., weekly inventory of ambulance)	Paramedic
Primary medical care provider for occupational workers in support of operational activities	Occupational health physician or physician assistant
Medical authority/lead of occupational medical program	Occupational health physician or physician assistant
Provides medical diagnosis of occupational medical injury or illness (including fitness for duty)	Occupational health physician or physician assistant
Provides medical readiness information of workers to support operations	Medical administrative staff
Maintains records showing fitness for duty to support operational activities	Medical administrative staff

a number of roles and responsibilities performed by the ES&H organization. Many corporations require monthly reporting of ES&H statistics and metrics. This information is used by executive management in evaluating company performance and can be used by the corporation in the proposal process for future work. The program management function will regularly perform this work activity, and the ES&H manager should be familiar with all expectations and requirements associated with such reporting.

7.7 Chemical Safety

In 1986 there were a significant number of changes and additions made to the CERCLA with the passage of the Superfund Amendments and Reauthorization Act (SARA). One of the biggest changes under SARA was the passage of the Emergency Planning and Community Right-to-Know Act (EPCRA), also known as SARA Title III. There are a number of regulatory requirements associated with SARA Title III, which deals primarily with chemical safety. Because of this legislation it is prudent to have personnel dedicated to the chemical safety function to ensure the company is compliantly operating. The function of the chemical safety group includes the following:

- Maintain a chemical inventory
- Management of material safety data sheets (MSDSs) associated with all chemical products
- Procurement process associated with chemicals
- Compliance to regulatory requirements associated with auditing and reporting of chemical products

According to both SARA Title III and OSHA, facilities are required to have material safety data sheets (MSDSs) for chemicals that are used at a work location. Personnel must be allowed access to all MSDSs for chemicals they are using, and the MSDSs must be made available to state and local officials and local fire departments. OSHA mandates that the MSDS is required to contain the following information (includes new requirements as of June 1, 2015):

- Identification of the product, manufacturer, address, and contact information
- Hazard identification, including all hazards of the chemical and required labeling
- Composition and information on the ingredients

- First aid measures that address symptoms and effects
- Firefighting measures such as suitable extinguisher, equipment, and chemicals from the fire
- Accidental release measures, including personal protective equipment (PPE) and cleanup methods
- Handling and storage precautions
- Exposure controls and personal protection, including permissible exposure limits (PELs), threshold limit values (TLVs), and appropriate engineering controls
- Physical and chemical properties that list the chemical's characteristics
- Stability and reactivity and possibility of hazardous reactions
- Toxicological information, including routes of exposure, related symptoms, acute and chronic effects
- Ecological information (required by agency other than OSHA)
- Disposal considerations (required by agency other than OSHA)
- Transport information (required by agency other than OSHA)
- Regulatory information (required by agency other than OSHA)

Regulations pertaining to what information is required on an MSDS continue to evolve, and as noted above, there is additional information required as of June 1, 2015 (included in above requirements). Depending upon the size of the facility, it may require more than one person to manage the MSDSs, required inspections of chemical inventory, required reporting of chemicals, and interfacing with local and state agencies to coordinate emergency response actions (as part of EPCRA). Because the liability for noncompliance exists across more than one regulatory authority (OSHA, Environmental Protection Agency (EPA), Department of Transportation), it is prudent for the ES&H manager to particularly stay abreast of chemical safety requirements to ensure the company is compliantly operating. There are a number of regulations that pertain to the procurement of chemicals. In particular, when purchasing chemicals the procurer should determine:

- Is the procurement of this chemical vital to the mission of the company?
- Can another chemical that is less hazardous, but still effective, be purchased?

As part of most environmental goals and sustainability programs reduction in the hazard and use of chemicals is required, and those companies that are successful in achieving such reductions are typically viewed as being environmentally friendly. In addition, because the requirements associated with chemical management are applicable across multiple regulatory agencies, the ES&H manager must understand all reporting requirements, including those that may be associated with local regulatory agencies.

7.8 Summary

Just as an ES&H organization is the backbone to performing work in a safe and compliant manner, the program support group of the ES&H organization serves as the cornerstone to program development and field implementation. The program support group is an integral part of the ES&H organization that is chartered to support the entire organization in developing and managing the support functions that are necessary to enable regulatory compliance while supporting the people aspects of the organization, operations, and company.

8

ES&H Training

8.1 Introduction

Safe, effective, and efficient ES&H performance is grounded by personnel who are trained and qualified to perform their job. When we refer to training in this section we are speaking of a process used to teach the needed knowledge and skills to successfully and safely perform work. Whether you are an executive in a corporation or working on an assembly line, training is an essential part of how successful you will be at your job or profession. In particular, the ES&H manager must be mindful of training that is required for personnel to perform their jobs in a safe and compliant manner and in compliance with regulatory requirements.

Training is fundamental and necessary to all work that is performed. All regulations require some form of training, and effectiveness of the training plays a large role in workers being able to successfully identify and mitigate hazards in the workplace and safe performance of work. It is also worth noting that training programs and records are frequently evaluated by regulators as part of routine inspections and accident investigations.

For example, 10 CFR Part 50, "Domestic Licensing of Production and Utilization Facilities," Section 50.120, "Training and Qualification of Nuclear Power Plant Personnel," requires that each nuclear power plant licensee or applicant for an operating license implement training and qualification programs that are derived from a *systems approach* to training. Paragraph 50.34(b)(6)(i) requires that an application for a license to operate a nuclear power plant include information concerning organizational structure, personnel qualifications, and related matters. Subpart D, "Applications," of 10 CFR Part 55, "Operators' Licenses," requires that operator license applications include information concerning an individual's education and experience and other related matters to power plants. The NRC Regulatory Guide 1.8 (2000) addresses requirements associated with qualification and training of personnel for nuclear power plants.

According to DOE Order 426.2, *Personnel Selection, Training, Qualification, and Certification Requirements for DOE Nuclear Facilities* (2013), companies must establish a process for the selection and assignment of personnel who are involved in the operation, maintenance, and technical support of Hazard Category 1, 2, and 3 nuclear facilities. The process (systematic approach) must include factors such as background, experience, education, and medical examination (as applicable), and should be based on the ability of the person to meet job performance requirements. Job performance requirements should be based on personnel demonstrating they have the requisite knowledge, skills, and abilities to properly perform work in accordance with the safety basis. The selection of personnel may involve a specific type of test to ensure the right candidate is selected.

Training requirements associated with the Environmental Protection Agency (EPA) focus on workplace activities designed to eliminate the release of pollution and waste into the environment and are contained within the respective regulations. These regulations also address asbestos, lead abatement, and other highly toxic or cancer-causing materials. Training requirements encompass not only the trainee but also the trainer. Examples include the following:

- Ozone-depleting substances and training required for motor vehicle air conditioning service technicians
- Asbestos abatement workers and required training for workers, inspectors, planners, project designers, contractors, and supervisors
- Hazard communication standard and affected personnel who work with chemicals
- Hazardous waste remediation sites and personnel who perform work at those sites
- Hazardous waste large and small quantity generators and training of personnel on how to manage and ship waste
- Lead-based paint abatement and required training for workers, dust sample technicians, inspectors, risk assessors, supervisors, and project designers

There is a large amount of training required by the regulations, and the ES&H manager is generally responsible to ensure that training being provided to workers within the company is occurring in compliance with the requirements. Please note that a person who is trained may not equate to being qualified. In some cases, a medical exam is required for workers to be considered qualified for a job function. Specific regulatory requirements will identify if additional actions are needed to be qualified beyond training.

Training for workers performing work in high-risk environments, including ES&H managers, professionals, and technicians, is generally required to follow a systematic approach to training (SAT). SAT is a method that provides a consistent approach to the establishment and implementation of performance-based training programs. An excellent publication that describes a logical SAT is the *DOE Training Program Handbook: A Systematic Approach to Training*, DOE-HDBK-1078-94 (DOE, 1994). A tailored version of the approach in DOE-HDBK-1078-94 is presented below and can be used to develop a training program.

8.2 Systematic Approach to Training

Training that is conducted efficiently and effectively and is directly related to the needs of the job (performance based) is fundamental to corporate and individual success and accomplishment (DOE, 1994). The SAT is a performance-based training program based on a five-phased approach to training and can be used by the ES&H manager in the development of a training program or material. Figure 8.1 (DOE, 1994) depicts the SAT process and is further described below.

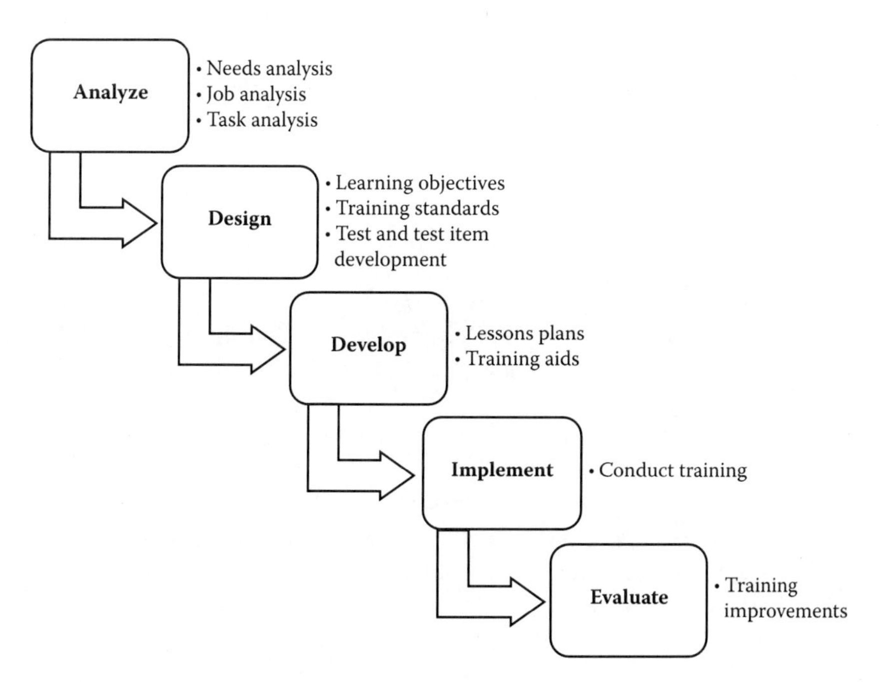

FIGURE 8.1
Systematic approach to training model.

8.2.1 Analysis of Training

The analysis of training requirements are identified by the following three components: a needs analysis, job analysis, and task analysis. These components represent methods for gathering information that are fundamental to training development. Below is a general summary of the three analysis components:

Needs analysis. The needs analysis is routinely used to identify solutions to job performance deficiencies. Prior to creating a new course or revising an existing training program, a needs analysis should be conducted to determine that training is the appropriate solution to the worker performance issue. Development of the needs analysis should address (DOE, 1994):

- Do performance deficiencies exist?
- Are employees capable of performing their jobs?
- Do they frequently perform the job?
- Have previous employees been able to perform these jobs?
- Are operating procedures adequate or have they significantly changed?
- Are identified deficiencies training related?

Results of the needs analysis should be documented as part of the training program file to ensure a history file exists for that particular training class. A needs analysis template is shown in Figure 8.2 (DOE, 1994).

Job analysis. A job analysis is performed to identify a detailed list of work activities or tasks for a specific job or position. Job analyses can also be used to compare existing training programs (i.e., fall protection) to established requirements and identify deficiencies in the adequacy of program content. For the ES&H manager who is developing new programs or processes, the job analysis provides an appropriate forum for documenting the information necessary to identify tasks associated with the job. It is a good idea to consult the regulations to ensure training is completed based on the requirements.

All relevant information regarding position-specific job analyses should be documented in a job analysis report, which becomes part of the training program file for each specified position. A good job analysis enables the person performing the analysis to more fully understand the nature of the job, identify how much of the job analysis work has already been completed, and identify information needed to write a job description if one does not already exist. A job analysis template is shown in Figure 8.3.

Requester: ____________	**Unique ID:** ____________
	Date Issued: ____________

1. Work activity to be improved:

2. Identify reason for why improvement is required:

3. Frequency of the work activity:

4. Impact of not performing the work activity as defined:

5. Recommendations for improving work activity:

6. Develop training path forward or action:

____________	____________
(Trainer)	**(Date)**
Reviewed By:	
____________	____________
(Manager)	**(Date)**
____________	____________
(Requester/Supervisor)	**(Date)**

FIGURE 8.2
Needs analysis template.

Task analysis. The task analysis is used to define the knowledge, skills, and abilities required for safe and efficient accomplishment of a task. As training material is designed and developed for tasks selected for training, task analyses supplement the development of training and provide information related to such items as prerequisites for task performance and criteria for acceptable performance, such as limits, timeframes to perform task, critical steps, and additional information required to perform particular elements of the task or overall task, such as from a procedure (DOE, 1994).

When a task is complex or large, completion of a task analysis can assist in defining data needed to avoid overlooking key knowledge and skill requirements. Documentation collected during a task analysis should be retained as part of the training course file and periodically updated to ensure the records are auditable.

Job Location:__________ Analysis Lead:__________ Verified By:______________ Date:_______	Task Complexity in Table: Difficulty Importance Frequency			
Work Activity	D (1–5)	I (1–5)	F (1–5)	**Comments**

FIGURE 8.3
Job analysis template.

8.2.2 Design of Training

The design of training programs is based upon job-related/performance-based information originating from the analysis phase of the training. The basis for the design is the development of terminal learning objectives, training and evaluation standards, and developing the test and test items. Terminal learning objectives clearly state the measurable performance the trainee will be able to demonstrate at the conclusion of training, including conditions and standards of performance (DOE, 1994). They are directly linked to the task statement, and provide the framework by which development of training standards, enabling objectives, and lesson plans are conducted. When developing terminal learning objectives, the trainer should consider the following:

- Appropriate training setting
- Sequence of the terminal objectives

There is no wrong way to develop terminal learning objectives, but they should be logical and directly linked to the desired outcome of the training.

Training and evaluations standards provide the basis for the development of objective-based training materials and maintain consistency in the evaluation of student performance. Each training and evaluation standard is directly related to a specific job task (or group of very similar tasks) identified during the job analysis and should be reflective of affected equipment or material. The training section contains the task title and number, the terminal and enabling objectives, and applicable references. Information contained in this section forms the basis for subsequent training development activities.

The evaluation section may contain a performance test that includes prerequisites, amplifying conditions and standards, and instructions to the trainee and the trainer. The performance test is used to measure the adequacy of a trainee's performance on a particular job-related task (DOE, 1994). It is during development of testing and evaluation standards that the majority of task analysis occurs, since many of the knowledge and skill requirements for task elements are identified while writing these standards. In addition, during the development of testing and evaluation standards it is recommended that the subject matter experts (SMEs) be consulted for input. Development of tests and test questions is focused on measuring trainee performance against criteria stated in the learning objective. Considerations when developing test questions should include the following:

- Layout and logic of the test questions
- The number of test questions to be included and time allotted
- Expected knowledge
- Validation of contents of test questions
- Incorporation of questions into a test database for future use

Tests are a form of evaluation that instructors use to measure the results or effectiveness of their stated objectives. Test questions should be developed and evaluated in an objective nonbiased, rather than subjective biased, manner. Figure 8.4 is an example training evaluation standards worksheet.

8.2.3 Development of Training

The development of training programs is written to reflect job-related information generated as part of the analysis and design phases (all phases and actions preceding training development). The development of training programs should consider the following:

- Training methods to be employed
- Lesson plan development
- Training aids needed to enhance learning

Instructions – Check each item after review:

- ☐ 1. Does the training criterion evaluate how adequately the trainee performs the work activity?
- ☐ 2. Does each training criterion evaluate the work activity for each required step?
- ☐ 3. Does each training criterion evaluate how to perform each work activity given operational costs and schedule?
- ☐ 4. Does each training criterion meet the standard of adequate work?
- ☐ 5. Have specific instructions for administration and evaluation of the training criterion been provided?
- ☐ 6. Does each training criterion represent a target number of the sample population?
- ☐ 7. Does the training criterion require that the work activity task be "performed" in those cases where performance is suitable and feasible?

FIGURE 8.4
Training evaluation standards review checklist example.

Once developed, a proficient training program allows personnel to comprehend and understand defined learning objectives relative to their job function. Table 8.1 lists different types of training methods along with an example of how that type of method can be applied in the ES&H environment.

Adequate lesson plans are a fundamental element of the training program and expand on information needed to communicate learning objectives. Lesson plans also ensure consistency in the material presented and should include learning objectives, content, learning activities, training equipment, and training materials needed for training and provide guidance for their use by the trainer.

8.2.4 Implementation of Training

Once training has been identified and analyzed, designed, and developed, one of the most rewarding tasks is to teach the training. As part of the process in delivering training, the trainer should consider the following:

- Training preparation. Lesson plans should be prepared and reviewed by the instructor prior to attempting to teach the class. The best instructors are those that have already thought of the tough questions (through real-life experience) and are prepared when the questions are asked.

TABLE 8.1

Training Method and Training Application Crosswalk

Training Method	Training Application
Classroom lecture	Used to communicate fundamental principles and academics. Examples include training on radiation protection fundamentals, Occupational Safety and Health Administration (OSHA) general industry standards, and environmental regulations.
Workplace demonstration	Used to train personnel on actual work practices used in the field. On-the-job training is a type of workplace demonstration. Examples include instrumentation calibration training, inspections of the workplace, and collecting an environmental sample.
Exercise role-playing	Used to communicate expected responses to workplace situations. Examples include simulator and emergency response training along with topics that require a high degree of human interaction (i.e., employee concern management).
Workplace walk-throughs	Used to enhance training that is conducted in the work environment. Examples include training on hazards identification process and identifying industrial safety noncompliances.
Self-pacing	Used when implementing a self-study. Examples include procedure revisions and communicating fundamental principles and practices.

- Delivery of the lesson. When teaching the class, the instructor should ensure all learning objectives are covered and understood. The best training classes are those where the trainer is able to actively engage the trainees and incorporate real-life examples.
- Evaluation of trainee performance. The primary purpose of evaluating training is to determine whether the trainee understood the training material and to provide feedback to the trainer as to how well the training was communicated. As with all methods for evaluating training, whether it is conducted using written tests or in the field, evaluation of training should include feedback to the trainees.

8.2.5 Training Evaluation for Effectiveness

To understand how effective the training was in being understood and retained, data should be collected, evaluated, and fed back into the training program to continually improve training effectiveness. There are a number of methods used to evaluate training effectiveness; below are a few suggested areas for consideration:

Field observations. There is no better method for evaluating the effectiveness and retention of training than observing how work is conducted in the field. This method is based on direct observation of personnel and can also include coaching by the instructor to remind workers of the appropriate method to perform work. This method of training evaluation is of great benefit to both the trainer and the individual worker. Information gained from observing work in the field serves as a means to improve future training.

Employee and supervisor feedback. Some of the best information used to improve training classes and material is received from employees and supervisors who have attended the training. They are on the front lines of daily work and have the best understanding of how useful and realistic the training is to the trainees. Not only is their feedback useful when received as part of training class review, but also the trainer should consider periodically going out into the field to better understand and keep abreast of how work is really being conducted so that experience and information can be incorporated into the training presentation.

Facility inspection and evaluation reports. Information received from inspection and evaluation reports provides an indication as to how work is being conducted, overall company health, and a mechanism for company or process weaknesses to be identified and fed back into the appropriate training classes. This type of method for evaluating training effectiveness is very useful when conducting training classes associated with regulatory compliance and on-the-job-training.

8.3 ES&H Training

Training for the ES&H manager or professional will vary depending upon the discipline (i.e., industrial safety, environmental, and radiation protection) and job position (i.e., manager, supervisor, professional, technician). Historically, training programs were designed focusing on a particular function; however, as companies are trying to reduce costs and improve efficiencies, the number of functions a professional may be performing as an SME tends to be more than one. Therefore, training should encompass several areas of responsibilities. Because of the large number of certifications that can be required for ES&H professionals to be viewed as qualified and competent, it is recommended that the ES&H manager have a good understanding and a strategy addressing certification maintenance and costs. Table 8.2 provides a generic list of training for the ES&H managers and supervisors/professionals.

8.4 Tracking of Training

Once training is completed, management of the data becomes of significant importance. Ensuring the training is conducted has no meaning if there is

TABLE 8.2
Generic List of ES&H Training Classes

ES&H Position	Example Class
ES&H manager	Leadership
	Personnel management and interaction
	Communication fundamentals
	Project management
	Cost/budget management system
	Emergency response
	Environmental/safety and health/radiation protection law
	Risk management
	Lead auditor
	Record keeping and case management
	Substance abuse
	Cyber security
	Classes relevant to disciplines and workers being managed
	Classes relevant to maintain professional license(s)
ES&H supervisor/professional	Supervisor leadership training
	Emergency response
	Behavior-based safety
	Risk management
	Assessment fundamentals
	Lockout/tagout
	Hazard communication
	Environmental/safety and health/radiation protection law
	Record keeping and case management
	Cardiopulmonary resuscitation
	Respiratory protection
	Medical exam
	Classes relevant to disciplines and workers being managed
	Classes relevant to maintain professional license(s)

not a convenient and efficient manner to retrieve proof of training completion. This becomes critical during inspections, evaluations, or during an accident investigation. This is also of importance to the ES&H manager because of the number of certifications that are maintained and the need to demonstrate continuous training credits to maintain licenses and certifications. Below is a list of items for consideration if the ES&H manager is involved with the selection of a training database or learning management system:

- Demonstrate continuous training credits needed to maintain licenses and certifications
- Scheduling of training

- Archiving and data management of completed training
- Tracking of costs associated with training, which could include training development, instruction, and employee attendance
- Maintaining professional license certification points and crosswalk to training classes that support continuing education points
- Reports and metrics that would easily notify management of missed, overdue, and upcoming training of personnel within their groups
- Access control of the database and security
- Personnel who are trained to operate the database

There are a number of off-the-shelf software applications that are available for management of training data. It is recommended the end user research the software programs and actually demonstrate the programs within his or her company to understand whether the software would be useful and adequate.

8.5 Training Records Retention

Most regulatory agencies require records to be retained for a defined specificity after the employee leaves the company, including training records. The periodicity varies depending upon the law, type of training class conducted, and job position. It is recommended the ES&H manager understand the specific requirements for both written records and electronic records, and that these requirements have been institutionalized through company policies or procedures.

8.6 Summary

Safe, effective, and efficient operational performance is grounded by personnel who are trained and qualified to perform their job. The ES&H manager must be mindful of training that is required for personnel to perform their jobs in a safe manner. Essentially all ES&H regulations require some form of training, and how effective the training is in enabling workers to successfully identify and mitigate hazards in the workplace and ensure work is conducted in accordance with regulatory requirements is frequently evaluated by regulators as part of routine inspections and accident investigations. Once training is completed, then management of the data becomes of significant importance. There are a number of off-the-shelf software applications that are available for management of training data. It is recommended the end user research the software programs and actually demonstrate the programs within his or her company to understand whether the software would be useful and effective. In addition, it is recommended the ES&H manager understand what are the specific requirements for training records maintenance.

9

Continuous Improvement of Environment Safety and Health

9.1 Introduction

One of the many challenges in today's work environment is the need to continue to seek efficiencies in the way we deliver and perform business functions, from the perspective of continuously improving both existing processes and efficiencies in cost and schedule delivery. The ES&H function is one of support—an integral component of the overall structure that produces a quality product and meets schedule and performance commitments. It is the responsibility of the ES&H manager and professionals to lead and identify methods for continuously improving their programs and field execution. This chapter addresses two methods by which improvement can be achieved:

- Assessing the organization to identify actions that will improve operational performance and costs
- Evaluating how the organization is performing by the tracking of issues and looking for trends that, if remediated, could achieve improved performance and costs

9.2 Assessing the ES&H Organization

A primary means of identifying areas for improvement is through the use of assessments. The term *assessment* refers to a systematic method for evaluating people, processes, and equipment in order to determine whether defined procedures, protocol, and operations are being conducted in accordance with requirements and specifications and to identify opportunities for continuous improvement. The use of assessments is driven by both regulatory requirements and companies wanting to gauge how well operations are being conducted, how proficient management is, and whether the company is meeting regulatory and contractual requirements.

9.2.1 Assessment Techniques

Assessments are executed using a variety of techniques to ensure that a proper review is conducted. There are primarily three techniques used when performing assessments:

- Documentation reviews
- Interviews
- Field observations

Documentation reviews include evaluating procedures and technical documents that are used to establish and implement program requirements. Review of documentation should include verifying that they are appropriately implemented, correctly flow down regulatory and contractual requirements, and are of sufficient detail to reduce company risk from noncompliance—when executed. Documentation reviews are the first step of the assessment process and establish the foundation the assessor will use to determine whether interviews and field observations demonstrate consistency in implementation of the requirements.

Interviews should be conducted with personnel that represent part or all work activities within the company, including management, supervisors, and workers. Interviews should be conducted in a setting that allows for an exchange of information individually (more private) or in a group setting, whichever is deemed most appropriate in gaining access to the needed information. It is important when conducting interviews to document the dynamics of the interview, to ask open-ended questions that promote openness to communicate, and to note any feedback received from the interviewees.

An abundance of useful information can be obtained during the interview process; therefore, it is important to conduct the interview in an environment that is open, honest, and respectful. The interviewer must be mindful not to inadvertently provide or lead the interviewee to the answer when asking questions (this can occur when the interviewer is intimately familiar with the interviewee or company); otherwise, the interview data could be viewed as biased and incomplete.

Field observations are conducted by observing and evaluating whether work is being conducted in accordance with program and field procedures. Observations are also conducted to determine if employees are conducting work consistent with their training or recognized skill of craft. This assessment technique is extremely valuable in determining the overall culture of the work environment (i.e., procedure compliance, roles and responsibilities, etc.).

Additionally, workplace surveys can be used when there is a desire is to understand the perceptions of a larger number of the work population. When a survey is used, whether electronically or manually implemented, the survey itself should be technically defensible and unbiased. The survey should contain the following attributes (Alston, 2013):

- Questions that are easily understood and can be interpreted by each employee in essentially the same manner
- Written in a manner that does not lead the worker to a particular response
- Questions that are clear and consistently stated

There are many types of assessments performed by organizations, such as independent assessments, management assessments, audits, surveillances, and specialty assessments. Many of these assessment types may be defined by the environment by which the company operates and is regulated. Managing and executing an assessment program that utilizes more than one assessment type can result in costs that were not originally planned and not recognized as adding value to the bottom line. As a means to still achieve continuous improvement in a cost-effective manner, the performance-based assessment model and techniques presented in this book have been proven to meet all the needs of the various assessment types, but at a fraction of the cost and with improved effectiveness.

9.2.2 Performance-Based Assessment Model

The traditional performance-based assessment focuses on the adequacy of the process that produces a product or service and then the product itself. For example, a traditional performance-based assessment would evaluate a piece of equipment or mechanical component on an assembly line to ensure it is being assembled in the correct fashion and determine whether the product performs as designed. This method has been adequate in assessing traditional manufacturing industries; however, the types of products and processes that exist in industry today are often unique or other than commercially standard, and do not readily adapt to traditional methods of performance-based assessments. Therefore, a new type of performance-based assessment model is defined in this book and presented in Figure 9.1 (Millikin and Vonweber, 2005).

9.2.3 Assessment Functional Elements

The performance-based assessment model presented in Figure 9.1 is comprised of three functional elements: compliance, effectiveness, and quality. Each of these elements is further described below.

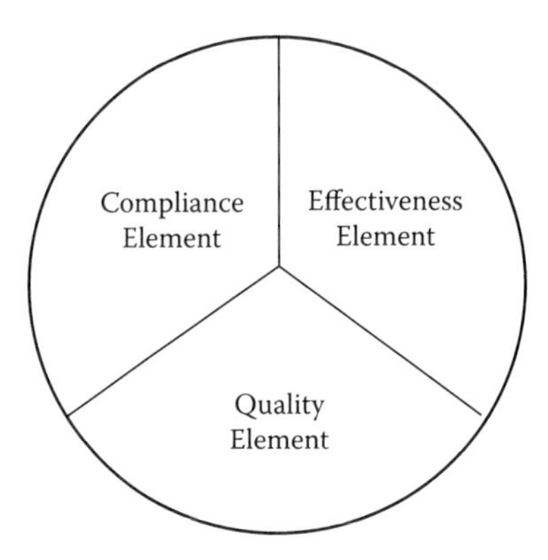

FIGURE 9.1
Performance-based assessment.

9.2.3.1 Compliance Functional Element

The compliance functional element focuses on verifying compliance with requirements through review of documentation and through observing requirements being executed in the field. Implementation of the compliance functional element begins with the determination of the regulatory and contractual requirements (where applicable) binding the assessed organization, function, or work evolution. The assessor then verifies that requirements are flowed down through implementing documents such as regulatory permits, procedures, or program plans, and then implementation of these requirements is verified in the field through direct observation and personnel interviews.

In safety, along with industrial hygiene and environmental, many requirements are found in the manufacturer's specifications for instrumentation, regulations, or permits issued by a regulatory agency. The criteria or lines of inquiry for the compliance functional element is qualitative versus quantitative. Traditional audits and surveillance programs are predominantly based on compliance-only criteria.

9.2.3.2 Effectiveness Functional Element

The effectiveness functional element uses a similar approach to the compliance functional element, such as seeking to determine implementation of requirements and compliance to procedures, but also addresses how well people, processes, and equipment perform to nonregulatory criteria. Examples of nonregulatory criteria include the following:

- Functionality: The product should meet the needs of the customer and company expectations.
- Roles and responsibilities: People understand their roles and responsibilities.

- Management span of control: Personnel are managed consistent with the company's policies, values, and mission.
- Work performance: Work is planned and performed in an efficient manner.

The criteria or lines of inquiry for this functional element are qualitative versus quantitative. The final determination as to whether a process or people are proficient in the work being assessed is subjective. Assessments that are effectiveness driven are traditionally used in evaluating management systems.

9.2.3.3 Quality Functional Element

The quality functional element focuses on whether a particular process or equipment conforms to design or model specifications and may be simplistic in criteria used to define acceptability. Most companies have quality assurance documents that define specific criteria or attributes that processes and equipment must meet. The criteria may originate from a national consensus standard or manufacturing specifications. They may be both qualitative and quantitative in nature and definitive with an acceptability determination. Assessments that are primarily quality driven are the traditional assessments performed by quality assurance personnel.

The performance-based assessment model utilizes all three functional elements by assigning a percentage that each element will contribute to the overall assessment through its criteria/lines of inquiry or assessment questions. The addition of the three percentages sums up to 100% of the three functional topical elements. The weighing of the assessment model components—compliance, effectiveness, and quality—will drive the outcome of the assessment as to whether it will be focused on a regulatory, effectiveness, or quality-related endpoint objective.

For example, if the objective of the assessment is to determine compliance to an Occupational Safety and Health Administration (OSHA) regulatory subpart, then 70% of the assessment criteria/questions would be related to specific compliance or contractual requirements, 20% of the criteria would be related to effectiveness of the people, program, process, or system, and 10% would be related to the quality of the process or product in meeting predefined specifications. More simplistic, if the assessor wants 10 questions total for lines of inquiry or criteria, then 7 of the 10 would be related to the OSHA subpart, 2 would be efficiency related (i.e., how are personnel documenting field observations), and 1 would relate to quality (i.e., are the materials being used for controlling the hazards to meet specified standards).

A second example of the performance-based assessment may be the evaluation of a management system that was established to identify and control hazards during execution of work. Although the hazards identification and control process is established to meet regulatory requirements, how well the system is working and proficient at mitigating the hazards would be the focus of the assessment. Therefore, 20% (2 out of 10) of the assessment criteria or lines of inquiry would be based on compliance to the requirements, 60% (6 out of 10) would be based upon criteria that would assist the assessor in determining how well the process or system is functioning, and then 20% (2 out of 10) would be related to the quality of the process or system in meeting quality criteria of the program. The assessor can use any number of combinations of criteria or lines of inquiry; the examples provided above were presented to simplify to the reader the quantitative approach for developing and executing a performance-based assessment.

It should be noted that the frequency by which ES&H managers and professionals are required to assess their programs may vary depending upon regulatory or contractual requirements. Once the purpose and focus of the assessment is defined, the assessor then develops an assessment plan to document and communicate assessment approach, methods, and defined criteria lines of inquiry or questions.

9.2.4 Performance-Based Assessment Plan

The assessment plan provides an outline of the assessment and is the primary means of documenting how the assessment will be conducted. The assessment plan may be a formal written document consisting of several pages, or a very short one-page summary. The assessment plan is written approximately a month prior to field performance of the performance-based assessment. Within the assessment plan objectives, schedule of performance, identification of the assessor(s), and lines of inquiry or questions are documented and communicated to the assessed function or group.

9.2.4.1 Assessment Plan Objectives

Objectives for the assessment may be as simple as one sentence that defines the endpoint that the assessor is determining, or may consist of an overall objective and additional criteria that collectively can be used by the assessor in determining whether the endpoint of the assessment is met. For example, a simple assessment may consist of the following objective:

> The company is compliantly and effectively implementing 29 CFR 1910.217, "Mechanical Power Presses."

A more in-depth assessment may consist of the following objective:

> Determine whether the company is compliantly and effectively implementing 20 CFR 1910.217, "Mechanical Power Presses," by demonstrating the following criteria:

- Machine components shall be designed, secured, or covered to minimize hazards.
- Electrical components of the equipment are inherently safe.
- Safeguards of the equipment are in compliance with OSHA requirements.

The depth to which the assessment objective is defined is largely driven by the end use of the assessment results. Specifically, if the assessment will be used by the company in determining general performance of the company in a specific area, then a simple assessment objective may be sufficient. However, if the company is going to use the assessment results in defending its performance against regulatory or contractual issues, then it is suggested that more in-depth assessment objectives be defined.

9.2.4.2 Schedule of Performance

The schedule of performance as defined in the assessment plan is driven by work/calendar dates. It is recommended dates be defined for preparatory activities, field activities, and assessment documentation. Depending upon the purpose of the assessment, this information can be extremely important to those organizations or the work groups being assessed. Feedback from the assessed organization or work group may be needed to ensure assessment performance can be conducted within the defined timeframe. A unique identification number should also be assigned to the assessment for tracking purposes, along with the location where the assessment is to be performed.

9.2.4.3 Identification of Assessor(s)

The lead assessor and the assessment team member should possess the skills needed to successfully complete the assessment. The lead assessor has some specific requirements that include the following:

- Selecting team members with the appropriate skills and knowledge based on the objective of the assessment
- Developing the assessment plan
- Ensuring that the all elements of the scope are assessed
- Serving as the focal point for responding to technical issues and disagreements
- Validating completeness of the report before presenting to the customer
- Presenting the assessment results to the customer

The assessors are responsible for complying with the assessment scope and supporting the team lead in completing the assessment. It is also recommended that if the team members are comprised of personnel external to the company (i.e., industry subject matter experts), then this information is highlighted in the report to further demonstrate independence and add credibility to the assessment team. If the assessment is anticipated to be used in the future to defend the company or work evolution, often resumes or brief biographies of assessment team members are attached to the final assessment document.

9.2.4.4 Assessment Criteria and Lines of Inquiry

Assessment criteria and lines of inquiry (LOIs) are comprised of either discrete statements or questions that the assessor will be asking in the field to meet the assessment objective and determine whether the company is compliantly and in a quality manner meeting the assessment objective(s). Results of the criteria or LOIs are used to identify the following:

- Practices that are noteworthy and positive for the company
- Observations that if corrected could improve company and work performance
- Issues that will be required to be corrected to meet defined regulatory, performance, and quality standards

Example lines of inquiry for an assessment on mechanical power presses are presented in Table 9.1.

9.2.5 Assessment Reporting and Results

Once the field portion of the assessment is complete an assessment report is generated. The assessment report is clearly written and organized to communicate best practices, observations, opportunities for improvement, and findings. Minimum information that should be included in the assessment report is listed below.

9.2.5.1 General Assessment Information

The unique assessment title and number should be identified in the report for future tracking purposes. In addition, the targeted location of the assessment, along with all locations evaluated, should be identified, including period of performance. In addition, all assessors that were used in the assessment should be identified. Due to schedule constraints of team members, there may be adjustments to personnel participating in the assessment that need to be identified.

TABLE 9.1

Mechanical Power Presses Lines of Inquiry

Assessment Line of Inquiry	Performance-Based Functional Element
Are the machine components designed, secured, or covered to minimize hazards caused by breakage, loosening and falling, or release of mechanical energy (i.e., broken springs)?	Compliance based
Is the brake capacity sufficient to stop the motion of the slide quickly and capable of holding the slide and its attachments at any point in its travel?	Compliance based
Is the motor start button protected against accidental operation?	Compliance based
Is the air controlling equipment protected against foreign material and water entering the pneumatic system of the press?	Compliance based
Does the equipment provide and ensure the usage of point of operation guards?	Compliance based
Does every point of operation guard meet specified design, construction, application, and adjustment requirements as defined in 29 CFR 1910.217?	Compliance based
Have operators of the equipment been adequately trained, and do they understand their responsibilities for how to operate the equipment?	Effectiveness based
Do supervisors understand the company responsibilities for ensuring the equipment is operated in accordance with regulatory and manufacturer's requirements?	Effectiveness based
Does the equipment perform as specified in the manufacturer's specifications?	Quality based
Are products generated from using this equipment meeting performance standards and expectations of the company?	Quality based

9.2.5.2 Signature Approval Page

All personnel who participated as team members of the assessment need to sign the approval page. This is important because it demonstrates the following fundamental principles:

- Concurrence by all assessment personnel that they agree that the issues identified are important enough that they need to be addressed
- Credibility to the assessed organization as to the quality and cohesiveness of the assessment team

9.2.5.3 Summary of Assessment Techniques and Observations

Documents reviewed, personnel interviewed, and field observations are documented. A good rule of thumb is not to list names of personnel evaluated and interviewed (for purposes of open communication and anonymity); list only position titles. This is important for establishing trust, in particular when the person being interviewed has requested anonymity or when working with groups of workers. Results of the assessment, including logic, used to determine whether the assessment criteria/lines of inquiry were met are to be documented. When issues are identified they should be categorized into one of four types:

- Finding: A finding is an issue that has been identified as a direct violation of a standard, procedure, or requirement.
- Observation: An observation is an issue that has been identified that although it is not a direct violation of a standard, procedure, or requirement, there could be potentially negative impacts to the project or company if left uncorrected.
- Opportunity for improvement: An opportunity for improvement is an issue that has been identified that although it may not result in or have the potential to result in a noncompliance, performance of the project or company would be improved if executed.
- Best practice: A best practice is an activity performed that results in a positive benefit to the project or company—and could be of benefit to other areas within the project/company or could be used by other companies as well.

As part of the assessment report, all findings, observations, opportunities for improvement, and best practices should be documented and summarized in the executive summary of the report. It is also a best practice to review the assessment results and conclusions with the work group that was assessed to minimize negative feedback prior to issuing the final document.

9.3 Performance Metrics and Performance Indicators

Performance metrics and performance indicators are used by ES&H managers, both within ES&H and across the company, to evaluate and monitor performance. In many instances managers use the terms *performance metrics* and *performance indicators* interchangeably; however, they are distinctly different. The difference between the two will be discussed in further detail later in this section.

9.3.1 Performance Metrics

A performance metric is a direct measurement of something with a defined goal and performance that is monitored over time. The term *performance*

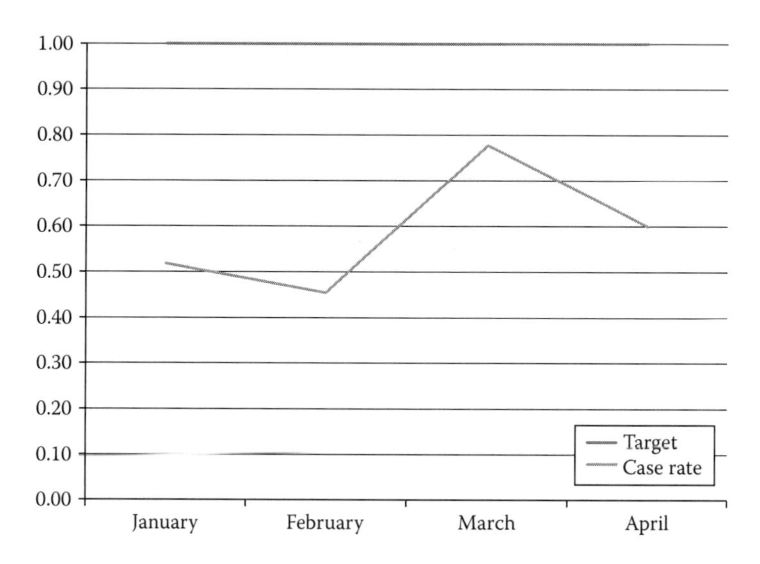

FIGURE 9.2
Total recordable case rate performance metric.

metric refers to a quantitative or qualitative measure by which operations, performance, and effectiveness can be monitored and analyzed most commonly against a standard or goal. For example, a common performance metric that is used in safety and health is total recordable case (TRC) rate. An example of a typical TRC metric is presented in Figure 9.2.

Based on the metric presented in Figure 9.2, the company's TRC is better than the target goal of 1.0. This is a simple performance metric; however, it is indicative of determining how well a company is performing in terms of protecting personnel from industrial hazards and proficiency of the hazards identification and control process. Other examples of performance metrics that are used by ES&H professionals and management include the following:

- Number of near misses
- Number of first aid occurrences
- Number of monthly assessments performed
- Number of environmental spills
- Number of radiological skin contaminations

Additional example metrics are presented in Figures 9.3 and 9.4.

9.3.2 Performance Indicators

A performance indicator (PI) is a qualitative analysis of several metrics combined or may be a single PI that is qualitatively based. The term *performance*

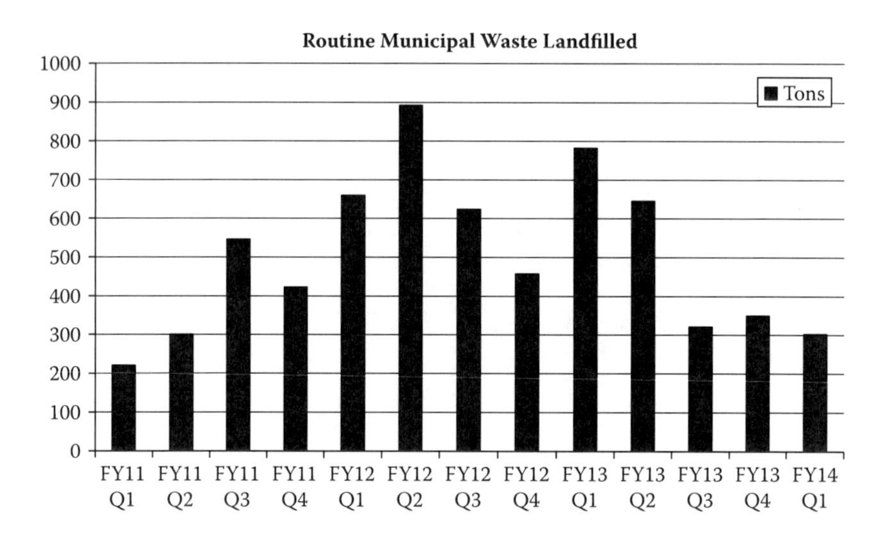

FIGURE 9.3
Waste disposal metric.

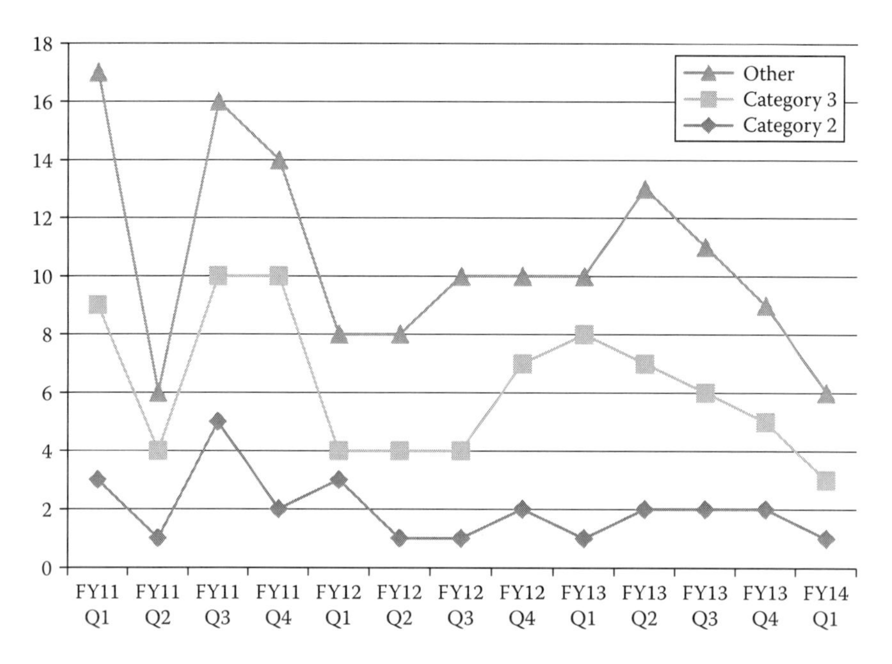

FIGURE 9.4
Environmental releases metric.

TABLE 9.2

Safety and Health Program Performance Indicator

Safety and Health Program—March				Overall Trend ↑
S&H Performance Metric	**Monthly Goal**	**January ↑**	**February ↑**	**March ↑**
TRC	<1.0	0.52	0.48	0.79
Days away and restricted time	0	1	1	1
Number of first aid cases	0	5	3	6
Near misses	10	2	6	2
Number of safety observations	30	0	6	5

indicator refers to a qualitative means by which a company goal or objective will be or is likely to be met. A PI is also monitored over time, and the determination of performance is more subjective than a performance metric. As depicted in Table 9.2, an example of a PI would be overall performance of the safety and health program.

Evaluation of the data presented in Figure 9.2 indicates that although the TRC is below the goal, the overall safety performance is trending in a negative direction. This PI is indicating that should performance continue in a negative direction, a more serious accident may occur. This is a simple PI and commonly used; however, it is indicative of determining the health of a company's safety and health program and is useful in communicating positive and negative trends with management and others within the company.

A key performance indicator (KPI) is an indicator that is more important than other PIs, and is primarily related to cost, schedule, or performance of key company systems. KPIs are comprised of a set of indicators that are used to gauge or compare company performances in terms of meeting their strategic and operational goals. Companies have fewer KPIs than PIs, and they are strategically defined and oftentimes used to indicate overall company performance related to scope, cost, schedule, productivity, and safety. In many cases, profit (or a specified percentage of award fee) is based upon meeting these KPIs; therefore, they are of particular interest to the executive management team. Examples of KPIs are listed below:

- Number of lost workday cases
- Number of radiological events
- Number of mechanical parts rejected based on not meeting quality assurance (QA) criteria
- Zero regulatory fines
- Number of personnel leaving the company

The use of performance metrics and indicators is an integral part of the continuous improvement process and can be a very useful tool for the ESH manager when identifying areas for improved performance, whether it is personnel or systems related. There are two types of metrics and indicators used by ES&H managers to monitor program execution: leading and lagging.

9.3.3 Leading Metrics and Indicators

A leading metric or indicator is a representation of data, which over time, can provide a measure or indication of positive or negative performance related to people, processes, or the overall company. It is a measurable factor that changes before performance can start to follow a pattern or trend. Most often leading metrics are used to provide an early warning of potentially declining performance.

Leading metrics and indicators can be used to identify risks and negative performance before an accident or some other negative consequence occurs. A good example of a leading metric or indicator for ES&H is the number of near misses per month. This information can then be superimposed on the operational or production schedule to identify work conditions that, if left unmitigated, could potentially result in a more serious incident, such as a first aid or recordable case. Below are specific examples of leading metrics and indicators used by companies:

- Near misses (definition of near misses must be clearly defined)
- Behavior-based observations (that can be monitored by the type of behavior observed)
- Training or medical appointments attended
- Communications specifically related to topical areas
- Number of self-identified issues (can also be viewed as a lagging indicator)
- Number of issues/nondeficiencies documented against regulatory subpart(s)

An increase in the specificity of the leading metric or indicator results in higher quality trending information.

9.3.4 Lagging Metrics and Indicators

A lagging metric or indicator is a representation of data that over time can provide a measure or indication of whether the historical performance meets goals and objectives for that particular topical area. Lagging indicators use historical information that measures past performance. Although lagging indicators are not generally used to modify performance behaviors and systems prior to an actual incident, they are extremely useful to gain an overall understanding of an organization or company's performance and to identify if interim corrective actions are needed to improve performance. In addition, they are used by management to identify long-term corrective actions needed to ensure specific goals, objectives, and improved performance are achievable.

The majority of metrics and indicators used in business today are typically lagging indicators. A good example of a lagging metric is the total recordable case (TRC) rate used to determine safety performance. Additional examples of lagging indicators include the following:

- Radiological contamination events
- Environmental spills
- Regulatory violations
- Vehicle incidents
- Employee concerns/human resource issues

9.3.5 Qualities of Solid Performance Metrics and Indicators

There are characteristics that ES&H managers and professionals should consider when developing performance metrics and indicators:

- The metric should be measurable and quantitative if possible. Otherwise, the results and meaning of the metric will be subjective to debate.
- The metric or indicator should be meaningful and relevant to work or focus areas of concern for the company. A lot of time is wasted reviewing indicators that really do not add value, but rather are developed for personal interests and curiosity. It is suggested they should correlate to performance or profit targets.
- The metric or indicator should not be susceptible to manipulation. More often than not, management and some personnel may choose to modify the data to show positive performance rather than the true performance. The ES&H manager needs to be cautious and ensure that he or she fully understands each indicator and how the data is generated and to be interpreted to prevent potential improper manipulation of it.
- The metric and indicator should be based on available data. There are times when the desire is to create an indicator to evaluate performance through qualitative methods. Be cautious because data should be quantitative and readily available; otherwise, the purpose of using performance indicators will be lost and they will not be meaningful for improving processes and performance within the company.

- The metric or indicator should exhibit the potential to be integrated with other metrics and indicators, such that they can be used to not only evaluate performance, but also identify other areas for improvement that would support the primary indicator.

9.3.6 Performance Metrics and Indicators in the Continuous Improvement Process

The use of performance metrics and indicators in today's business environment is extremely critical in monitoring performance and identifying opportunities for continuous improvement. Because of the competitive nature of business and the ever-increasing need to be cost-effective, performance metrics and indicators allow ES&H managers and professionals, line management, and executive management to be proactive in managing processes and systems and take actions that can improve performance. The integration and use of the performance metrics and indicator improvement model (PMIIM) is depicted in Figure 9.5. Each of the specific actions related to the PMIIM are further described below:

1. Establish performance metrics and indicators. Establishment of the correct performance metrics and indicators is one of the most critical elements of the PMIIM. The selection of the correct topical areas to measure can be used to monitor the link to strategic decisions on

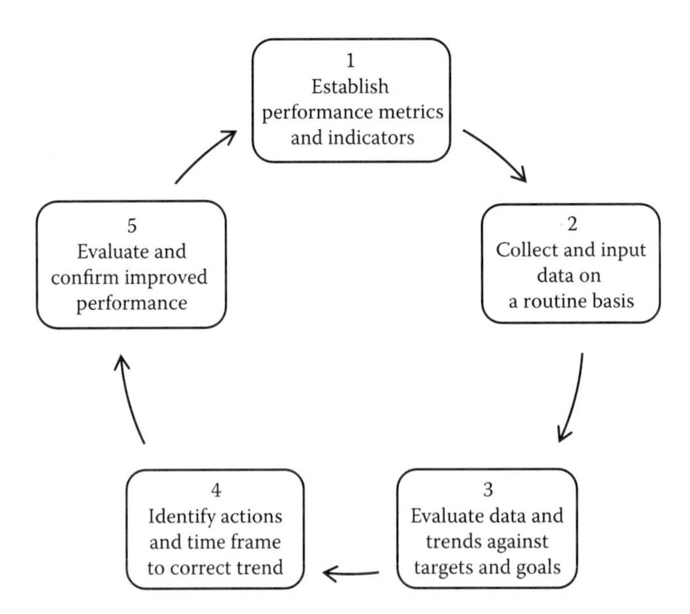

FIGURE 9.5
Performance metrics and indicators improvement model.

how to manage a company, function, or work group. When selecting the performance metrics and indicators to evaluate the ES&H manager should consider the following questions:

- What are the goals and objectives of the company within the next 6 months and year?
- What are some of the key liability issues that are facing the company, function, or work group?
- Within a specific functional area of ES&H, what are areas that have shown specific issues over the past year?
- Are there any specific corrective actions in place for a topical area that the company should be monitoring?

When making the decision on the metrics and indicators it is recommended the ES&H manager solicit input from his or her peers and company management to ensure the right areas are being targeted for continuous improvement.

2. Collect and input data. Once the metrics and indicators have been identified, the next step in the PMIIM is to collect data. This task requires input from multiple sources, including the following:
   - Operations
   - Safety and health professionals
   - Environmental professionals
   - Radiological control professionals
   - Multiple sources, such as databases used to track radiological dose, safety, and environmental observations

   It is important that all individuals who will be responsible for collecting the data understand their role and how the information collected will be used as input for the metrics and indicators.

3. Evaluate data and trends. Evaluation of the data generally begins with inputting it into a spreadsheet or database and generating a graph. If evaluating data as part of a metric, the data should be analyzed against predetermined goals or standards. Performance is measured on a specified timeframe (typically monthly) and evaluated to determine if a trend exists. The definition of *trend* is a pattern of events, incidents, items, activities, processes, or causes reflected by other processes or operational data that is unacceptable or undesirable and is important to the degree that corrective action or close monitoring is deemed appropriate by management. The term *trending* includes adverse trend, emerging trend, and monitoring trend and is intended to include other patterns of concerns, such as not yet categorized trends or areas of improvements. Trending is further defined below:

- Monitoring trend: A monitoring trend does not warrant corrective action, but needs to be evaluated to ensure that it does not evolve into an emerging trend or adverse trend. A monitoring trend may be used to track previously identified adverse or emerging trends to closure.
- Emerging trend: An emerging trend exists if the trend does not meet the criteria for an adverse trend but represents performance for which action(s) need to be taken to ensure it does not result in an adverse trend.
- Adverse trend: An adverse trend is the recurring occurrences of a problem or adverse condition that involves similar tasks and causes that are significant in nature or are critical to the success of the project as determined by management (e.g., significant issue designation). Adverse trends include programmatic or systemic conditions, or significance category, including programmatic or systemic conditions that have the potential to negatively impact the company's mission and objectives.

The intent of the trending process is designed to identify potential areas of weakness before there is a negative impact on operations or productivity. This deviation can occur in a positive or negative direction. There are many different approaches to quantitative analysis of data in determining whether a trend exists; significant information has been published on various methods of trending. A basic rule of thumb is a consistent 3-month time period of deviation that is above the predetermined goal. Whatever method is used in determining a trend, it must be technically defensible and easily explained. In addition to the quantitative analysis of the data, there may also be other factors that are not directly related to the performance indicators, but could influence and impact performance indicator deviations. For example, a recent layoff of personnel may result in a negative trend in the TRC rate.

4. Identify actions to correct a trend. Upon identification of a trend there are primarily three types of corrective actions:
    - Compensatory corrective actions: Actions that prevent the issue from immediately reoccurring: stop gap measure. Compensatory actions are short term and intended to be temporary until long-term preventive actions are complete for a trend, issue, or problem. For example, a compensatory measure for preventing electrical equipment from being operated until permanent engineering controls can be established would include issuing a standing order that requires everyone to understand what can

happen should the equipment be inadvertently operated without the proper controls.

- Remedial corrective actions: Actions that remedy the immediate trend, issue, or problem. For example, a remedial measure associated with preventing electrical equipment from being operated until permanent engineering controls can be established could include installing a padlock on a particular piece of equipment, thereby disabling individual operation of the equipment.
- Preventive corrective actions: Actions that prevent or reduce the probability that a trend or incident can recur. For example, a preventive measure for preventing electrical equipment from being operated may include installation of permanent engineering controls, such as an interlock.

5. Improve performance. Upon identification and execution of corrective actions, the metric or indicator in need of improvement should be monitored to determine whether the corrective actions were adequate or whether additional actions may be warranted. It is important the management and responsible organizations assume ownership of their specific metrics and indicators because they should be used by them to better manage their specific groups or organizations. Also, identified metrics and indicators should be reevaluated on a periodic basis to ensure they are still providing the necessary information needed to better manage the company, function, or group, and whether the goal or standard that has been previously established is still appropriate.

9.3.7 Use of ES&H Metrics and Performance Indicators in Company Management

Performance metrics and indicators are an extremely dynamic tool for the ES&H manager to identify and target specific areas for improvement, but they are also a good communication tool for operations and executive management. Performance indicators are depicted on charts that represent a number of issues, incidents, and events over time (monthly, weekly, yearly). Many companies schedule monthly performance indicator review meetings with management to review performance and identify areas for improvement. In addition, performance indicator review boards may be established at designated workplace locations so that employees, as well as management, can review their performance in relationship to established goals. In some cases, these review boards can be managed by a designated group of workers and managers to further instill ownership of performance and success of the company.

9.4 Summary

Continuous improvement of the ES&H function is paramount to improving efficiency and effectiveness of processes in every company. Management must ultimately own the continuous improvement processes, but often it is the ES&H manager who plays a vital role in coordinating and implementing the process. Two key elements of a continuous improvement process are performance-based assessments and performance metrics and indicators.

The assessment program is essential in the identification of areas for improvement. A performance-based assessment model is presented that integrates three functional elements into a cohesive and cost-effective method for performing an assessment. Performance metrics and indicators are critical for monitoring performance and identifying positive or negative trends. A performance metrics and improvement model is presented that defines how and when performance metrics and indicators can be proficient in managing and monitoring improved company performance. By using the tools presented in this chapter, the ES&H manager and professional will ensure positive program performance and minimization of risk.

10

Project Management Approach to Environment Safety and Health

10.1 Introduction

Oftentimes projects within an ES&H organization are started but not completed because of a lack of structure and focus. Utilizing project management to approach these projects can optimize the ability of the ES&H manager to get projects accomplished and increase organizational efficiency. Project management is simply a process of planning, organizing, controlling, and directing resources for a defined period of time in order to complete a defined goal. The project management approach to ES&H management is a proactive and progressive way to manage the diversity of tasks and activities that are required to be performed in the effort to protect the worker, the environment, and the community.

10.2 The Project Management Approach

Project management is a great process to use to successfully plan, manage, control, and execute projects successfully. The concept is a proven effective process that can be used in a diverse business environment. Project management concepts are also useful in the management of projects in an ES&H organization. When using project management concepts all of the phases of a project are critical and must be addressed in order as listed in Figure 10.1.

When using project management skills to guide the operational parameters of an ES&H organization and projects, it may be necessary for the ES&H manager to use a scale-down approach or modify one or more of the project management cycles to better fit smaller projects and functional tasks. The stages of a project and the associated tasks that may be performed during each stage are listed in Table 10.1. Also found in Table 10.1 are the types of tasks that may be expected to be performed in each stage.

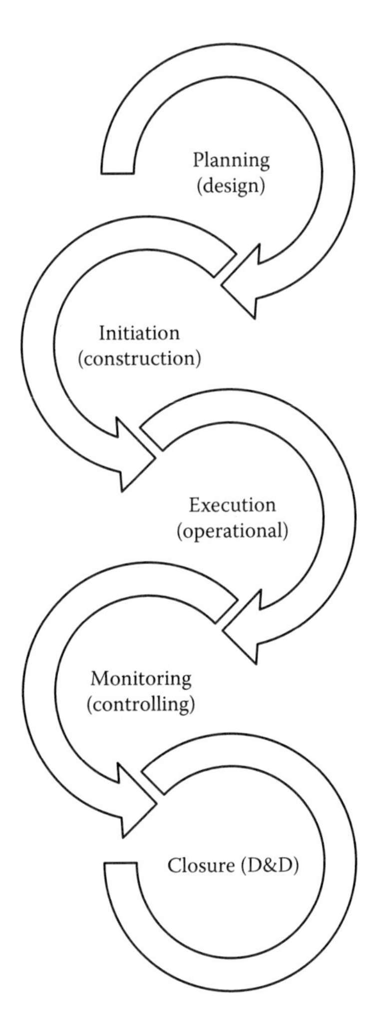

FIGURE 10.1
ES&H project management approach.

10.3 ES&H Scheduling

Scheduling is an integral part of projects and functional organizations as it provides the means to transform the project from a vision to a well-thought-out plan, and it provides the means to track progress, resource needs, and expenditures. The project schedule is the tool that communicates the activities that need to be performed, the timeframe for completing the activities, and the resources needed to complete each activity. A project or an ES&H schedule can range from simple to complex. Because of the uncertainty

TABLE 10.1

Project Stages and Task

Stage	Task
Planning	• Develop the project scope. • Determine the resource needed (staffing needs, skills, software, hardware). • Develop budget, including any subelements of the budget that may be associated with a work breakdown structure. • Develop communication and risk mitigation strategy. • Begin development of a project/work plan. • Develop the project schedule.
Initiation	• Justify and determine the goal for the project. • Name a project manager. • Determine key team members. • Determine and define roles and responsibilities for key team members. • Identify the resource needs, the cost estimates, and a timeline. Realize that these parameters may change since it is early in the project.
Execution	• Ensure project team has the required knowledge (training, briefings, etc.). • Implement project plan and schedule. • Direct and lead the project team. • Communicate key aspects of execution of the work. • Interface with management, workforce, and stakeholders.
Monitoring	• Measure and monitor project performance against the project plan. • Ensure project progresses according to the plan. Modify plan as needed. • Manage project risks and challenges. • Conduct status review meetings. • Document lessons learned.
Closure	• Get feedback for project usability and overall management of the project. • Resolve all newly discovered issues. • Document all lessons learned. • Ensure closing documents are complete and stored for future retrieval if needed.

involved in project planning and implementation, the schedule should be reviewed regularly, and revised as needed. A schedule may continue to evolve throughout the project as it progresses. Benefits of using a schedule include the following:

- Monitors and controls project activities
- Tracks project progress
- Evaluates the impact of delays on project completion (project risk)
- Provides a clear pictorial of where resources are needed and the availability of resources

When developing a schedule there are certain guidelines that should be followed to produce an adequate schedule regardless of the size or complexity of the project. These guidelines are listed below for consideration:

- Identify all of the major activities.
- Identify timelines that include start and ending dates.
- The schedule should coincide with the work breakdown structure depicting the sequencing of work. The work breakdown structure defines and divides the scope of the project into manageable discrete activities so that they can be easily understood.
- Identify the exact sequence of work to be performed.

In some cases the completion of some work activities may be dependent upon the completion of other preceding activities. This correlation should be defined through a linking of the related activities on the project schedule. A pictogram of the elements that should be included in a schedule is shown in Figure 10.2 with the preferred order of sequence. Note that each step is dependent upon the preceding step. If the preceding step is not clearly defined, then the step that follows may not be clearly defined. Errors in any of the critical steps listed can impact the project greatly.

Tables 10.2 and 10.3 provide examples of how to use the project schedule approach in managing an ES&H organization and projects. A comprehensive schedule allows the ES&H manager to communicate the complete effort

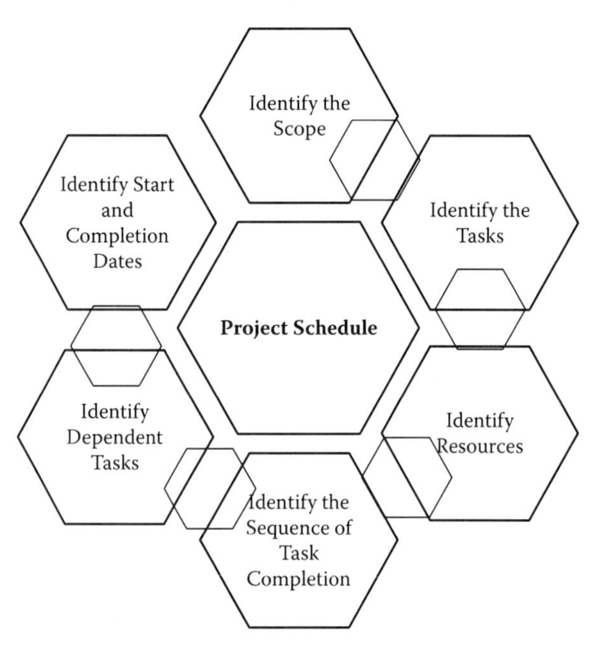

FIGURE 10.2
Schedule sequencing.

TABLE 10.2

Environment Safety and Health Organization Schedule—Example

ID	WBS	Manager	Group	Task Name	Start	Finish	Percent Complete	Resource Name	Strategic Plan Code
				ES&H leadership training					
				Determine content of training					
				Develop script for training					
				Convert script into PowerPoint					
				Convert script and PowerPoint into articulate software and record the course narration					
				Perform alpha and beta testing of course					
				Management self-assessment					
				Develop lines of inquiry					
				Assemble the assessment team					

Continued

TABLE 10.2 (Continued)

Environment Safety and Health Organization Schedule—Example

ID	WBS	Manager	Group	Task Name	Start	Finish	Percent Complete	Resource Name	Strategic Plan Code
				Develop the assessment strategy					
				Review lines of inquiry with the assessment team					
				Conduct assessment					
				Complete and distribute report					
				Fall protection program					
				Review the OSHA fall protection standard					
				Determine if the ES&H fall protection procedure is in line with OSHA					
				If changes are necessary present and gain approval of changes to the operations council					
				Revise procedure					

TABLE 10.3

Hearing Conservation Program Upgrade Project—Example

ID	WBS	Task Name	Resource Name	Percent Complete	Start	Finish
		Identify at-risk population				
		Develop a schedule to fit test high-risk population				
		Develop hearing conservation fit test cards				
		Complete fit testing of identified population				
		Identify hearing protection custom-fit vendor				
		Identify custom vendors and associated cost				
		Select, negotiate, and sign contract with selected vendor				
		Schedule high-risk population needing custom-fit hearing protection				
		Schedule custom-fit testing				
		Develop a schedule to fit test the rest of the population				
		Revise hearing conservation training to include fit testing protocol				
		Conduct one-on-one training on the proper method to use for inserting hearing protection devices				
		Request legal interpretation of mandatory participation in hearing conservation program				
		Revise hearing conservation procedure to include requirement for hearing protection fit testing (include frequency of testing and groups to be fit tested and referral process)				

of the scope of work performed by the functional area to include projects, in terms of cost, resources, and the timeframe necessary to perform the scope assigned to the functional area. Scheduling creates order and helps ensure work is conducted within time and budget constraints. Components of an ES&H functional area schedule can include the task identification (ID), work breakdown structure, manager responsible for the task, functional group, task name, start and finish dates, percent completion, and resource/labor category that will complete the task. If the schedule activity is in support of the organization strategic plan, then it is a good idea to tie the strategic plan number to the task. Oftentimes there is a separate scheduling organization that can produce a schedule formatted from professional scheduling software. The examples presented in Tables 10.2 and 10.3 are simple and can be used by an ES&H manager if no scheduling program is available.

10.4 Managing Cost

In preparing the budget, all of the associated costs are estimated and then totaled to develop a budget for the functional area that includes the projects that are expected to be performed. Each ES&H project and functional task should have an associated cost, whether it is the cost of the labor hours of a subject matter expert (SME) or computer programmer, or the cost of supplies and hardware. It is important to know that in order to maximize the chances of meeting the budget, the schedule must stay on course. The first step in cost management is to estimate the costs of each activity in the project. Costs include both human resource and physical resource costs. Because this step often occurs in the planning phase, it is important to understand that the estimated costs are a "best guess" at the actual costs of each activity. The next step is to create a realistic project budget. In this step, determine the cost baseline and the funding requirements for the project.

A full-blown financial plan may not be necessary when planning functional projects if the resources that are being used have salaries that are being paid as an overhead function within the organization. However, it would be a good idea to capture the cost to keep track of the complete cost of the project. This information can be used if there is a need to justify staffing levels for the current year as well as upcoming years.

10.5 ES&H Managers as Project Managers

It is common for ES&H managers to take advantage of using project management tools and concepts; however, it is necessary to ensure that the manager is equipped with the necessary skills. There are many tools available to aid managers in successfully managing projects to completion. Recognizing that no one tool will address all of the needs of the manager and the project, it is important for the ES&H managers to select the best tools that suit their skill level and the characteristics of the project. The managers in an ES&H organization have a multifaceted role. ES&H managers are most effective when they have some technical knowledge in their respective functional areas. They serve oftentimes as project manager for the projects that are performed within their functional areas. In their project manager role they are responsible for managing internal projects with various scopes, cost, and timelines. Therefore, it is advisable to provide project management training for ES&H managers in order to ensure that they have the skills and knowledge to manage a project. The skills that are also important in managing the functional area and projects include the following:

- Leadership skills
- Team-building skills
- Differing opinion resolution
- Technical knowledge
- Resource allocation skills
- Budget management skills
- Organizational skills
- Planning skills

In an ES&H organization it is not uncommon for the functional manager, as well as the senior and highly skilled professionals, to function in the project manager role, depending on the size and complexity of the project. Oftentimes preparing the professionals to function as a project manager on a small-scale project is used as a method to prepare them for other leadership roles with greater responsibilities. In such case, generally a skilled functional manager will oversee the professional while functioning in the role of project manager. This role presents some unique opportunities to facilitate the projects that are being worked since the functional manager has control of resources as well as ongoing commitments and responsibilities.

Project managers having the responsibility for completing a project within budget and schedule are somewhat at the mercy of the functional, line, or resource manager since these managers typically control resources. Generally, project managers do not have control of the resources. These project managers are successful when they have a good working relationship with the manager that is in direct control of the resources. The critical skills that are important for managers to succeed in their roles are discussed in more detail below.

TABLE 10.4

Project Management and ES&H Management Attributes for Success

Leadership Attributes for Project Managers	Leadership Attributes of ES&H Managers
Integrity	Integrity
Good communication skills	Good communication skills
Ability to inspire a shared vision	Ability to inspire a shared vision
Competent in the project subject area	Competent leader
Ability to delegate	Ability to delegate
Team builder	Team builder
Good problem-solving skills	Good problem-solving skills
Strategic thinker	Strategic thinker
Good organizational skills	Good organizational skills
Ability to maintain composure	Ability to maintain composure
Skilled at building relationships	Skilled at building relationships
	Change agent
	Specific educational training in ES&H
	Skilled at building relationships

10.5.1 Leadership Skill

A leader is someone that possesses a high level of integrity and competence and can inspire people to willingly follow. Table 10.4 shows the leadership skills needed for ES&H managers that are required to perform well in their role as managers. Also found in Table 10.4 are the attributes required by functional managers to be viewed as competent and able to inspire the organization to achieve the level of performance needed.

In viewing the attributes above for project managers and ES&H managers, it is clear that the leadership attributes of both roles are primarily the same. Therefore, it is conceivable that in a functional organization a good ES&H manager can also function as a competent project manager.

10.5.2 Team Building

There is a difference between the role of a functional team and a project team. A functional team is a permanent team that has been established to conduct activities within a particular organization, group, or department. On the other hand, a project team is assembled to complete a defined scope with a defined start and completion date. At the completion of the scope the team will dissolve and move on to another project or back to the functional area. Assembling a cohesive project team in the functional area, comprised of mostly functional members, makes team building easier since the majority of the team members are accustomed to working together. This relationship

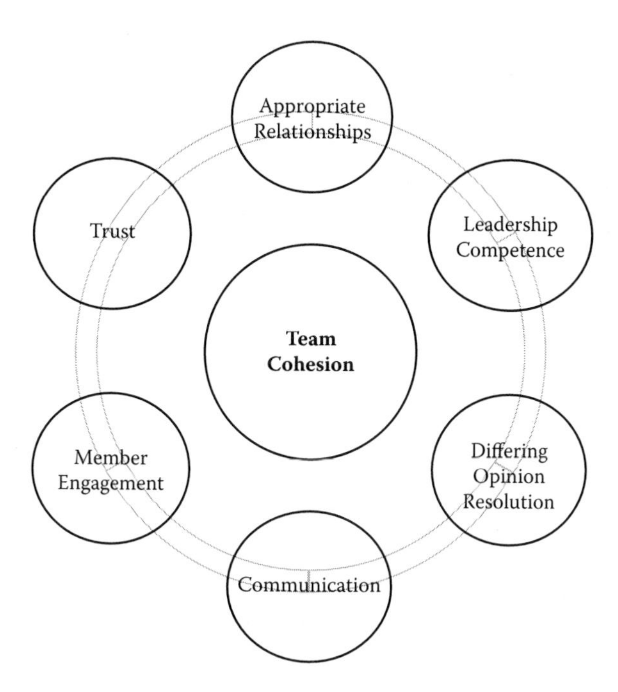

FIGURE 10.3
Team cohesion model.

makes it easier for the project manager to assist the new members when it comes to assembling into the team with minimal team-building activities.

In order for a team to be proficient, team cohesion is necessary. Team cohesion is the glue that keeps a team together and functioning as a group with shared goals, preventing the group from team fragmentation. Elements that should be focused on to aid in building team cohesion are listed in the team cohesion model shown in Figure 10.3.

10.5.3 Differing Opinion Resolution

Recognizing that conflict will occur, a skillful manager will establish a mechanism to handle conflict before it occurs. The differing opinion resolution process is a great tool to use and is considered part of the role of a project manager and the functional manager. The ES&H manager often deals with resolving differing opinions frequently because he or she is responsible for the technical aspect of the business as well as the people part of the business.

ES&H managers often interact with customers, internal and external. Therefore, handling conflict resulting from differing opinions is a skill of importance just as it is for a project manager. These managers should be better suited to handle conflict resulting from differing opinions, since dealing with conflicts of this sort is an ongoing functional responsibility.

10.5.4 Technical Knowledge

The ES&H manager generally does not possess all of the technical knowledge needed to complete functional tasks or projects. However, it is advisable that the manager does possess enough technical knowledge to understand the functional concepts and how to make decisions based on the information presented from other technical experts within the discipline. When a manager does not possess the technical knowledge required, generally decisions are made that appear to be without merit or purpose. Some level of technical knowledge is a must if project management concepts and tools are to be used in managing projects that the functional area embarked upon.

10.5.5 Resource Allocation and Management

Resource allocation for a project in the functional area where the ES&H manager is assuming the role of the project manager can make staffing the needs of the project smoother since the functional manager ultimately has control of the resources. The term *resources* refers to everything needed to complete a task or project and includes people, equipment, material, and money. In the event resources are needed from other functional areas outside of the project organization, the ES&H manager will have to negotiate support time for the external resources. However, for projects internal to an ES&H organization, generally the need for external resources is limited, with the exception of the need for computer program support and development.

10.5.6 Organizational Skills

The functional manager has an edge when functioning as the project manager since he or she has knowledge of how the functional organization is structured. Although organizational skills are of particular importance during the project design and start-up stage, these skills are very important from project conception to closure. Organizational skill is also important to the functional manager when developing and implementing a strategy that will support the needs of the business. The important elements that are key ingredients that help to develop organization skills are listed in Figure 10.4. These attributes are viewed as organization skill enhancers when applied together consistently.

10.5.7 Planning Skills

Planning skills, like organization skills, are necessary to move an organization forward and increase productivity. In order for a functional manager to lead a team or an organization successfully, he or she must possess the

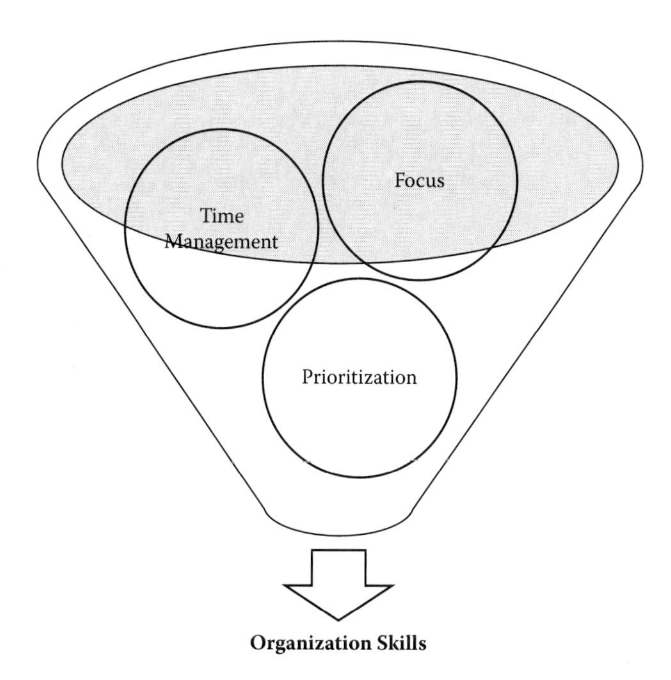

FIGURE 10.4
Organization skill enhancers.

ability to strategically plan for the longevity of the organization or the team. Likewise, a project manager must possess good planning skills. Without good planning skills it is unlikely that a manager can develop and monitor a project schedule or lead a project or functional team.

10.6 Time Management

Time management is critical, especially when tasks require competing resources to contribute to the completion of more than one project. Effective time management helps managers implement established strategic goals and determine objectives to reach those goals on schedule. Because there are only 24 hours in any given day and time cannot be created, the ability to manage time is critical. Below are activities that can help make time management easier to navigate.

If project or functional managers are unable to manage their time, then it is most likely that they will not be able to manage the team time, and this can place the project at risk. Each day the project/functional managers should review what activities or event took place that impacted the project. Effective time management can help make managing a project easier. There are specific elements that must be sharpened and kept in mind when the desire is to manage time for the functional area and the team. These elements are shown in the time management portal in Figure 10.5. Activities that can be used to aid in enhancing time management are included in Table 10.5.

TABLE 10.5

Time Management Enhancers

Activity	Benefits
Refrain from micromanaging.	Let the team selected do the job they were selected to do. This provides time for the project manager to concentrate on directing the project.
Develop the plan and schedule.	A plan will keep everyone focused on the milestones, therefore reducing wasted time that may result from distractions and confusion.
Have productive meetings.	Limit the amount of time wasted in meetings.
Project managers should not be involved in task or activity completion.	Project managers lose focus of the entire project when they get involved in completing work activities.
Use the schedule to depict status.	Avoid asking every project member to provide status during meetings. This can take a lot of time and does not add value. Use the project schedule for its intended purpose.
Develop an attention list.	Keeps focus on important activities that need to be addressed immediately.

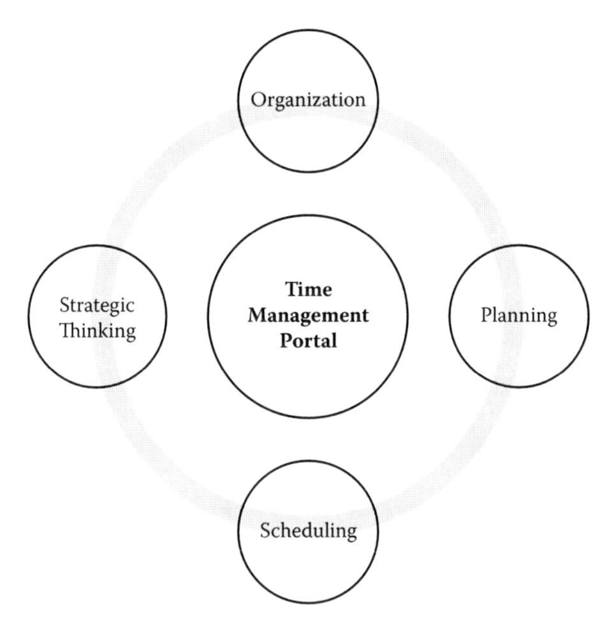

FIGURE 10.5
Time management portal.

10.7 Summary

In this chapter we have tried to make the case for integrating project management concepts into the strategy of an ES&H organization through the use of project management tools and concepts for all functional area projects and activities, where feasible. In addition, we have pointed out that the functional manager is required to have, and does possess in most cases, the same skills and characteristics needed by project managers and can easily function in the role of a project manager. Project management skills and tools are important attributes to help the functional manager efficiently manage his or her organization.

Project management is broadly defined as a systematic approach to planning, scheduling, and controlling work. Integrating project management tools and practices can help an organization to

- Facilitate completion of key projects
- Control cost associated with projects
- Provide a means to build project management skills of the management team and professionals
- Eliminate those functional projects that go on and on with no completion ever achieved
- Resolve issues that prevent the organization from achieving efficiencies and technical growth

The use of project management tools and concepts in organizing and controlling work in an ES&H organization has the ability to increase efficiency and the compliance posture of the organization. In addition, it allows the functional manager to control the work within the function, train SMEs on project management concepts and practices, and provide ease of assigning needed resources to projects while balancing the work of the functional area.

11

Succession Planning

11.1 Introduction

What is succession planning? Succession planning refers to the process and strategy used to identify and develop potential internal candidates for key positions within a company.

Key positions are defined as those positions necessary to ensure that the company maintains its competitive advantage and is positioned to sustain continued success in the marketplace. Key positions may include those associated with executive and senior leadership positions or highly skilled technical professionals that may be difficult to find qualified candidates to fill. Oftentimes, these key positions require specialized training, licenses, or an experience level that is not easy to replicate. An important aspect of succession planning is to create a suitable match between the company's future needs and the desired career path of individual employees. A well-developed succession planning process can also serve as a means to increase the retention of high-performing workers.

Through succession planning the value of the prospective candidate is shown since the company is willing to invest in his or her development and subsequently his or her professional future. Workers that are included in the company's succession planning process feel that they are valued and that they have a voice in ensuring the success of the company and their own professional development. Succession planning is an ongoing process that can be complicated since most companies today have streamlined their processes and staff to gain efficiencies in operations and increase their economic bottom line. Many companies are conducting business with significantly reduced staff that is being asked to take on more tasks and in many cases working longer hours. In such cases, these already overtaxed employees do not have the time required to train or gain knowledge in other aspects of the business. It is important for leaders to think outside of the box and be creative in order to develop and prepare workers to take on critical roles that may result from retirement, reduction in staff, or workers leaving the company for other opportunities.

The success of an ES&H program is predicated upon the ability to attract and sustain the staff with the appropriate skills and knowledge. Many ES&H professions seek advanced education and certification and are required to be knowledgeable in a host of regulatory requirements. In order to support the training and certification requirements, management must invest time and resources in each professional. Filling positions from within provides a means to retain the staff that the management team has invested time and resources in developing. Therefore, succession planning is a great way to assist in maintaining skilled professionals needed to implement a sound ES&H program within a company.

Most leaders view succession planning as a process used only to identify and develop internal staff to assume key senior leadership roles. A successful plan must not only address senior leadership positions, but also include those technical positions that are key and hard to recruit. In order to recruit and maintain the appropriate level of talent, the succession strategy must take into consideration the company's employee retention strategy.

11.2 Employee Retention Strategy

The purpose of a retention strategy is to ensure that the company is able to attract and retain qualified staff, which ensures successful, profitable operations, and continues to meet the mission of the company. Employee turnover can be costly in many ways, such as negative impact on morale for the employees remaining with the company and the cost associated with recruiting and onboarding new employees. In today's corporate environment it is not easy to retain skilled science, technology, and engineering professionals. Many companies do not have a strategy in place to render the work environment attractive to workers. The companies that are able to attract and retain top-notch staff recognize that they must have more than just a job to offer employees. Prospective candidates are looking at all elements of a company, from the leadership team to the company's benefit plan and advancement opportunities. Where there is no retention strategy highly skilled employees tend to "come and go," leaving companies vulnerable to not being able to achieve goals and objectives. Companies that do not have a solid plan to retain workers can become a training ground for other progressive companies. Some attributes that may render a company attractive to employees and may aid in the retention process include the following:

- Competitive salary
- Competitive company benefits
- Advancement opportunities

- Engaging and rewarding assignments
- Employee engagement in decision-making process
- Trustworthy leadership team
- Fair implementation of policies and procedures
- Open and honest communication among the leadership team and employees
- Competent leadership
- Mentorship program that targets leadership development
- Programs that address and promote a healthy work-life balance
- Employee recognition program
- Flexible work schedule when appropriate

Succession planning cannot become a reality if the company is having a difficult time retaining the skilled staff. In such cases, a retention strategy is key to the success of a company's succession plan. Companies that are unable to retain staff can be viewed as unstable and will likely have a difficult time attracting skilled staff.

11.3 The Role of Management in the Succession Planning Process

Many companies delegate succession planning as a responsibility of the human resource (HR) department or each individual group supervisor. The HR department does not necessarily have the ability to determine the needs of each group within a company, whereas supervision tends to focus on their individual group. The responsibility of ensuring the needs of the business are met is entrusted to the leadership team. Therefore, succession planning is the responsibility of the entire management team.

Acquiring and managing required resources represents one of the most important elements of the needs of a business. Each level of management must be aware of and support the succession planning strategy. The leadership team is responsible for

- Determining the positions that are critical to the company's success
- Desired skills that are needed for each position
- Ensuring that the succession strategy is sufficient to meet the needs of the company

A succession plan will fail if the leadership team is not integrally involved and does not demonstrate continual support for the plan.

11.4 Attributes of a Good Succession Plan

A good ES&H succession plan is designed with the needs of the business in mind and focuses on preparing potential successors to step into roles to ensure viability and success of the company. The plan should include at a minimum the following:

- A listing of key positions
- A strategy that provides the means to identify the right candidate and prepare him or her to fill key positions
- Critical competencies required for key positions
- A listing of positions that may not be filled internally

This plan should also include the option to fill key positions through the external hiring process in the event the position cannot be internally filled. In such cases, a recruitment process should be in place and a selection committee should be identified. In addition, the succession plan is most effective when linked to the organization's strategic plan to ensure plan completeness and success. Other potential approaches that may be considered include consolidation of roles and responsibilities, and reorganization.

Maintaining a highly skilled ES&H staff is not an easy task because of the knowledge and skills required to successfully function in the field. The typical ES&H professional has degrees in the areas of science, technology, engineering, and math (STEM), to include industrial hygiene, safety, occupational medicine, environmental science/engineering, and the health services field. Because of the highly skilled and technical nature of these jobs, it is generally difficult to ensure the appropriate professional is available to step into these critically needed positions.

In designing a succession plan to accommodate the diversity of skills, training, and experience required to support an ES&H program the following steps, at a minimum, should be included in the succession planning process for each position:

- Identify all key positions.
- Identify key competencies for each position.
- Identify internal candidates.
- Internal candidate communication.
- Conduct gap analysis comparing the needs of the position against the qualifications and skills of the selected candidates.
- Develop a training and development plan.
- Implement a training and development plan.
- Evaluate the succession plan annually.

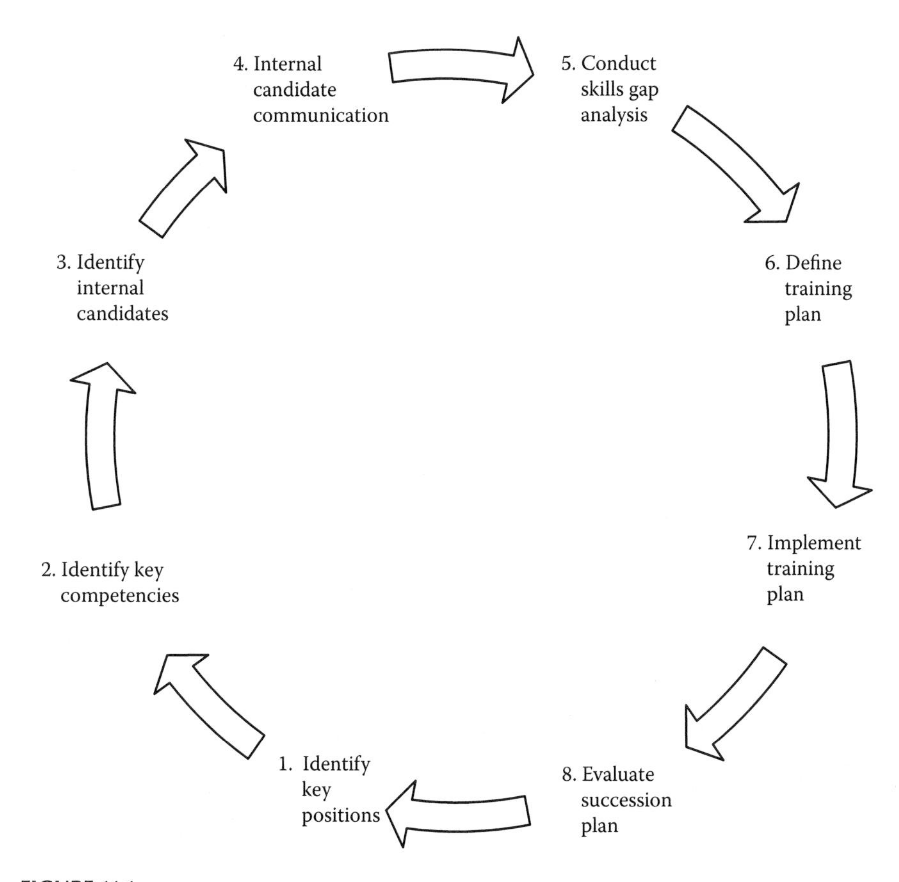

FIGURE 11.1
Succession strategy model.

The succession plan itself is an ongoing process that never ends, as indicated by the succession planning model pictured in Figure 11.1. The key to keeping the plan at the forefront of the leadership team is by evaluating the plan at least annually.

11.4.1 Identify Key Positions

The first step in the succession planning model is to identify all of the key positions. It is necessary to evaluate the goals and the scope of the organization to determine what positions are critical in delivering a successful and cost-effective ES&H program. Key positions are those positions that are associated with the scope that must be performed to ensure health and safety of the workforce, the environment, and to sustain the organization's compliance posture. Key positions in an ES&H organization include positions such as

- ES&H director/manager
- Occupational medicine director
- Radiation program manager
- Worker safety and health manager
- Analytical laboratory manager (if samples are analyzed internally)
- Environmental manager
- Assurance manager
- ES&H operations manager
- Injury and illness subject matter expert (SME)

Other positions can be considered key depending on the mission of the organization, organizational design and structure, and essential knowledge needed to sustain business success. Questions to ask to assist in identifying positions to include in the succession plan are

1. Does the position have significant impact on the day-to-day operation of the business?
2. Does the position have frequent interactions with customers, both internal and external?
3. Is the position responsible for developing the company's strategic plan?
4. Does the position require a unique skill set (e.g., security or emergency response background)?
5. Is the position critical to ensure regulatory compliance?
6. Is the position key to facilitating safe completion of work?

11.4.2 Identification of Key Competencies

The identification of key competencies includes defining the skills necessary to effectively carry out the tasks required by the key position. It is important to include all of the skills needed to successfully operate in the position, even if it has been determined that the proposed successor does not possess a particular skill set. Having the skills to perform successfully in a key position goes beyond having the technical skills and experience. In a management position it may not be as important to have the level of technical knowledge needed for the subject matter expert or technical professional. However, it is necessary that the soft skills necessary to gain trust and build relationships be included in the required competencies for management as well as technical professionals. Once the key competencies have been identified for each position it is a good idea to document the information on a critical skills worksheet, as shown in Table 11.1.

TABLE 11.1

Critical Skills Worksheet—Example

Position	Critical Skills
ES&H manager	• Leadership: Demonstrated skills in leading a staff of 100. • Budget management: Management of a budget of more than $50 million. • Technical knowledge: In at least three areas of ES&H (e.g., environmental, safety, and industrial hygiene).
Radiation safety manager	• Leadership: Management of a staff of 20. • Technical knowledge: Regulatory (varies depending upon regulatory agency) and field execution.
Worker safety and health manager	• Leadership: Management of a staff of 10–20. • Technical knowledge: Understanding of OSHA regulatory requirements and field execution/application.
Environmental manager	• Leadership: Management of a staff of 10–20. Experience in leading a small technical group of SMEs. • Technical knowledge: General knowledge in EPA, local, and state regulations and protocol.
Authority having jurisdiction—electrical, fire protection	• Technical knowledge: Electrical and NFPA requirements, life safety code compliance.
Biological safety manager	• Leadership: Experience in leading a small technical group of SMEs. • Technical knowledge: Centers for Disease Control and Prevention (CDC) regulations and protocol.
Industrial hygiene professional	• Leadership: Experience in leading a small technical group of SMEs. • Technical knowledge: Regulatory experience in OSHA and ACGIH and field execution.

11.4.3 Identification of Candidates

The first step in identifying potentially viable candidates involves looking across the organization at the existing talent pool. Evaluating the existing talent pool provides an indication of whether the positions can be filled from within or a search for an external candidate would be necessary. If the position can be filled from within, identify the candidates and the key position to be filled. It may be necessary to identify a backfill candidate in the event that the successor's position is also a key position. Upon identification of potential successors, then the selection(s) should be identified and documented. Table 11.2 provides an example template for how to document the internal selection process.

The selection process for an internal candidate is critical because the candidate has the potential to supervise or lead individuals that may have previously been colleagues/peers to existing staff. It is sometimes difficult for an internal candidate to readily adjust to a new position of leadership if he or she finds himself or herself managing a work group or department that he

TABLE 11.2

Example Successor Worksheet

Position	Critical Skills	Successor	Current Position
ES&H manager	Leadership: Demonstrated skills in leading a staff of 100. Budget management: Management of a budget of more than $50 million. Technical knowledge: In at least three areas of ES&H (e.g., environmental, safety, industrial hygiene).	Jane Doe	Key
Radiation safety manager	Technical knowledge: Radiation safety. Leadership: Management of a staff of 20.	John Doe	Nonkey
Worker safety and health manager	Technical knowledge: Understanding of OSHA regulatory requirements and field execution/application. Leadership: Management of a staff of 10–20.	Paul Doe	Nonkey
Environmental manager	Technical knowledge: General knowledge in EPA, local, and state regulations and protocol. Leadership: Experience in leading a small technical group of SMEs.	Pat Stone	Nonkey
Clean Air Act SME	Technical knowledge: EPA regulation—Clean Air Act.	Bob Stone	Nonkey
Authority having jurisdiction—electrical	Technical knowledge: Electrical and NFPA requirements.	Jim Stone	Nonkey
Biosafety manager	Technical knowledge: CDC regulations and protocol. Leadership: Experience in leading a small technical group of SMEs.	Shawn Doe	Nonkey
Industrial hygiene professional	Technical knowledge: Industrial hygiene.	April Stone	Nonkey

or she was once a part of. The ability of the internal candidate to transition into a leadership role must also be taken into consideration and be included in the succession strategy.

11.4.4 Candidate Communication

Once the candidates have been identified, management should immediately communicate to them the company's desire to include them as a part of the company's succession plan. The following information should be communicated:

- The selection process
- The role the candidate is being asked to potentially fill in the future
- The company's expectations for the candidate
- The preparation needed to prepare for the position

It is important to solicit feedback as to their acceptance of the succession plan being proposed so that both the company and the candidate agree they have the same goals and objectives. This is also a good time to discuss the employee's experience and begin gathering information that can be used to complete the gap analysis process. Oftentimes, workers may have additional experience and skills that have not been disclosed to the current management team. At the time the succession plan is disclosed to the candidate, a mentor should be assigned to aid the candidate in preparation for the new role that he or she is expected to take on in the future.

11.4.5 Gap Analysis

Upon mutual acceptance of the candidate succession plan, a gap analysis is developed to determine if the successor has the abilities needed to successfully function in the role or if there is a gap in training, skills, or knowledge that needs to be addressed. This type of analysis is commonly referred to as a skills gap analysis because it directly identifies various competencies lacking for each successor candidate. This is a very important step since it begins the preparation process for the new candidates.

The length of time and investment dollars needed to prepare a candidate is directly dependent upon the results of the gap analysis. The gap analysis must not only focus on the technical skills required, but also focus on the soft skills needed to further build trust and the relationships necessary to achieve success and implement the values and mission of the company. Listed below are some important elements to consider when developing a skills gap analysis:

1. Determine the skills needed to successfully perform the job scope in its entirety, including scope, cost, and schedule. Additional skills to be considered include communications, integrity, character, and fundamental leadership skills.
2. Rank the required skills in order of critical needs.
3. Review the skills with the proposed successor.
4. Compare the skills of the proposed successor with the skills needed for the critical position.
5. Document the gap in knowledge and skills and seek acceptance of the analysis with the proposed successor.

The information contained from conducting the gap analysis can be summarized in the matrix or worksheet shown in Table 11.3. The information contained in the example table represents the basic attributes of a skills gap analysis that is necessary to begin the development process for each candidate in preparation to fill a key position.

TABLE 11.3

Skills Gap Analysis Worksheet—Example

Position	Critical Skills	Successor	Delta in Skills
ES&H manager	Leadership: Demonstrated skills in leading a staff of 100. Budget management: Management of a budget of more than $50 million. Technical knowledge: In at least three areas of ES&H (e.g., environmental, safety, industrial hygiene).	Jane Doe	Lack of demonstrated leadership experience in leading technical organization Lack of demonstrated knowledge in one of the key subjects Lack of demonstrated knowledge in managing budgets over $10 million
Radiation safety manager	Technical knowledge: Radiation safety. Leadership: Management of a staff of 20.	John Doe	Knowledge in the area of radiation protection
Worker safety and health manager	Technical knowledge: Understanding of OSHA regulatory requirements and field execution/application. Leadership: Management of a staff of 10–20.	Paul Doe	Little knowledge in OSHA regulations
Environmental manager	Technical knowledge: General knowledge in EPA, local, and state regulations and protocol. Leadership: Experience in leading a small technical group of SMEs.	Pat Stone	Limited technical knowledge Limited leadership skills
Clean Air Act SME	Technical knowledge: EPA regulation—Clean Air Act.	Bob Stone	Technical knowledge acceptable
Authority having jurisdiction	Technical knowledge: Electrical and NFPA requirements.	Jim Stone	Technical knowledge acceptable
Biosafety manager	Technical knowledge: CDC regulations and protocol. Leadership: Experience in leading a small technical group of SMEs.	Shawn Doe	Technical knowledge and leadership skills acceptable
Industrial hygiene professional	Technical knowledge: Industrial hygiene.	April Stone	Technical knowledge acceptable

11.4.6 Define the Training and Development Plan

The results of the gap analysis should be incorporated into a training and development plan that charts the path and strategy for acquiring the required skills that were identified as being critical for the proposed successor. The training and development plan should be carefully constructed

TABLE 11.4

Training and Development Plan

Gap	Actions to Close Gaps	Completion Date
Leading a team of more than 100 professionals	Rotational assignment as worker safety and health team lead or some other positional assignment that will gain skills needed for managing large groups of personnel	
Lack of demonstrated leadership experience in leading a technical organization	Rotational assignment as environmental compliance team lead	
Coaching skills	Book: *Coaching*	
Knowledge in the area of radiation protection	Participate in a radiological assessment or potential loan assignments pertaining to radiological protection or attend formal training classes on select subjects	

to complement the candidate's current skills (Table 11.4)—with the goal of expanding those skills. Also included in the plan should be a strategy that is needed to ensure soft skills are acquired that are pertinent to achieving success. The training plan can contain a variety of mechanisms to facilitate the plan and close the gap in skills and knowledge. A training plan can consist of a combination of the following methods to enhance knowledge and experience:

- Formal classroom training (see Table 11.5)
- Mentoring and coaching
- Temporary short-term assignments (see Table 11.6)
- Reading assignments (see Table 11.7)

TABLE 11.5

Example Formal Training Worksheet

Getting the point across	Date scheduled:	Date completed:
Executive leadership	Date scheduled:	Date completed:
Radiation safety overview	Date scheduled:	Date completed:
Strategic planning	Date scheduled:	Date completed:
Corporate budget process	Date scheduled:	Date completed:
Public speaking and communication	Date scheduled:	Date completed:

TABLE 11.6
Rotational and Practical Assignment Schedule—Example

Six-month rotational assignment in the position of the worker safety and health manager	Date scheduled:	Date completed:
Participate in the internal ISO 18001 assessment team	Date scheduled:	Date completed:
Lead the Clean Air Act assessment team	Date scheduled:	Date completed:
Participate in a 10 CFR 835 review	Date scheduled:	Date completed:

TABLE 11.7
Reading Assignment Schedule—Example

Book: *Culture and Trust in Technology-Driven Organizations*	Date completed:
Book: *Dealing with Difficult People*	Date completed:
Book: *The 21 Irrefutable Laws of Leadership*	Date completed:
Book: *The Leadership Challenge: How to Make Extraordinary Things Happen in Organizations*	Date Completed:

The development plan should contain at a minimum the following:

- The skills that need to be acquired.
- The process used to acquire the necessary skills, for example, training or mentoring.
- The timeframe allotted to obtain each skill set. It is important to set completion goals with dates and hold the successor and mentor accountable.

A training plan can be written to be completed in a short time or for a long duration depending on the gaps being addressed. Table 11.5 depicts an example of a formal training worksheet that can be used to track scheduling and completion of needed training.

There are times when allowing the candidate to function in a rotational assignment to gain the experience and exposure needed is the best method to fill a gap. Table 11.6 depicts an example of a rotational and practical assignment schedule. If the assignment is maximized, this method can provide an effective means to increase knowledge and experience of the prospective candidate. However, this method requires the candidate to relinquish the current position and take on an assignment that is not permanent. This may make the candidate uneasy unless his or her position is retained for him or her or there is another opportunity to move into once the rotational assignment is complete.

Assigning the candidate books to read is a viable option to enhance knowledge. However, whether or not the delta in knowledge was filled may not be easily measured. In this case, the only way knowledge can be assessed is through the mentoring process. During mentoring sessions pertinent technical points can be discussed and the mentor can determine if the candidate has effectively learned the skill. The mentor must realize this is a subjective assessment of knowledge. To supplement the process questions can be developed to assist the mentor during conversation. Example questions may include the following:

1. What were the key points of the book or training material?
2. How can you apply the knowledge gained through reading the book or training material in your new role?
3. What was the most important point you took away from the book or training material?
4. What leadership skills will you take away from the book or training material?

Table 11.7 provides an example of a reading assignment schedule that can be used to track completion of the reading schedule.

11.4.7 Implementation of the Training and Development Plan

The training and development plan should be developed and personalized for each candidate with the appropriate timeline for completion. Special consideration should be given to avoid overwhelming the candidate with too many activities in a short period of time. Setting the expectation for completion in a short amount of time can hinder the candidate's learning process and may create some frustration in trying to complete the training and development plan due dates. The training and development plan should be reviewed frequently and progress tracked and assessed to ensure that the respective candidate will be prepared to take over and successfully perform in the proposed key role. Although the training and development plan originates from the gap analysis, it is important to recognize when modifications are needed that may include additional or less training.

11.4.8 Evaluate Succession Plan

The succession plan is reviewed and revised when there is a change in staffing, company business strategy, or otherwise annually. Succession planning is not a one-time event. In fact, it is an ongoing process and requires deliberate attention and planning in order to achieve the desired outcome. Succession planning is an essential process or tool that should be an integral part of the ES&H organization business strategies and practices. It must be repeated and modified as the mission and goals of the organization change, as well as every time there is a change in the succession candidates and structure of the organization. When developing a succession plan consideration should be given to whether a position can be internally or externally filled. Therefore, it is necessary to be familiar with the skills needed for each position and the skills and qualifications of the internal staff.

11.5 The Role of a Mentor in Succession Planning

What is mentoring? Mentoring is a structured transference or sharing of knowledge from one person to another for the express purpose of enhancing one's career. Many companies use mentoring as a means of preparing workers to take on new roles. The use of mentoring is becoming more popular for large companies as well as for small companies. Mentoring is used by organizations for reasons such as

- Preparing people to take on new roles or key positions
- Developing their pool of top talent, and succession planning

The role of a mentor is to guide and provide advice on how to approach and handle various issues that may arise. Therefore, it is important that the right mentor is selected and matched with the appropriate candidate. A mentor can be a line manager or someone external to the department or organization. In fact, at times it is more beneficial to have a mentor that is external to the mentee's organization. Mentoring should not be confused with training, and it is not, in most cases, the only actions to take to prepare successors. Mentoring provides an important opportunity to use internal talent to aid in training upcoming talent into various aspects of the organization culture. Successful mentors display good attributes, such as

- Good listening skills. In order to be a good listener, one must be focused on the words that are spoken. Good listening skills also require an open mind and the ability to provide undivided attention to the speaker.
- Ability to focus on the needs of the mentee. The mentor must focus on the needs of the mentee as opposed to his or her need or desire to gain experience in mentoring.

- Clear understanding of the mission, goals, and culture of the organization.
- Openness to providing constructive and honest feedback. A mentor must be willing to provide open and honest feedback to the mentee, recognizing that it is not always easy to provide feedback that is not viewed as complementary or positive. However, if the mentoring process is going to provide the means to close the gap in skills and performance, the mentee must view the process as open, honest, and nonthreatening.

11.6 External Hiring Process

There are times when there is a need to consider external candidates to fill key positions within an organization. If external hire is the path to be taken, ensure that each candidate is evaluated against a set of defined criteria to validate that the right candidate is selected. A critical fact to remember during the interview process is that not only are you interviewing the candidate to ensure that he or she possesses the attributes needed to function in the position, but also the candidate is using the interview process as an opportunity to evaluate the organization. When externally hiring, generally candidates are selected based on knowledge and experience. Therefore, there is not a need to develop an elaborate strategy or implement a plan to train and develop the new candidate in most instances. However, since external candidates lack knowledge of the company business strategy, culture, goal, and vision, time should be taken to assist them in becoming acquainted with the company and assimilation into the culture. One important factor to consider when choosing to fill a position with an external candidate is the means used to acclimate and integrate the new manager or technical professional with his or her new organization and the company. It is advisable to assign a coach to the candidate to facilitate integration into the organization.

11.7 ES&H Organizational Succession Strategy

Not only is succession planning necessary for technical workers to provide support to the customer in ensuring health and safety of the workers and the environment, but also it is necessary that a plan be put in place to ensure that the appropriate leadership team is recruited and maintained. ES&H leadership can be more complex than managing other aspects of the business, such as finance, procurement, engineering, or maintenance. The subjectivity and balance or priorities that are sometimes necessary to make good business decisions in the area of ES&H not only require a diverse knowledge in various regulatory requirements, but also require some level of decision-making proficiency and experience. Therefore, a manager is most effective if he or she has some knowledge or experience in the area. We are not suggesting that candidates without ES&H experience should not be considered as ES&H leaders. However, additional training, development, and support from leadership may be needed. Many of these managers fail to be proficient, and thereby are replaced by experienced and knowledgeable managers with some background in at least two facets of ES&H. The succession planning strategy may include grooming of internal candidates or recruitment of external candidates when needed.

The worker safety and health (WSH) functional manager is oftentimes viewed as the most difficult of all of the ES&H manager functional areas to replace. The fields of industrial hygiene and safety are included in the WSH area. Decisions made by the technical professional in the area of industrial hygienist (IH) are not always clear to most and historically have been based upon experience, which can appear to be subjective. Some of the decisions can be called subjective because, for example, every chemical or substance known to man does not have a regulatory/occupational exposure limit that one may be exposed to daily without adverse effect. The lack of an exposure limit may not necessarily mean that the chemical cannot or will not cause impact through prolonged exposure in excessive amounts or aggravate some preexisting condition. Therefore, the professional will need to take into consideration other attributes of the task, along with the products being used to determine safe work practices to prevent adverse impact to workers while working with the product.

ES&H professionals are generally highly educated and skilled in their ability to interpret regulations governing the field. Succession planning to backfill for these professionals when retirement is on the horizon or when a professional chooses to move on to begin a career with another organization requires a defined strategy and time to prepare successors. When selecting candidates the following pitfalls should be avoided:

- Avoid selecting candidates solely on their technical knowledge.
- Avoid selecting candidates solely because they have been with the company for a long period of time.
- Avoid selecting candidates solely because of relationships.
- Avoid selecting candidates that have a reputation for not being detail-oriented.
- Avoid selecting candidates that do not have a passion for safety and the environment.

- Avoid selecting candidates that are unable to assimilate data and make a good decision based on the data.
- Avoid selecting candidates that do not exhibit good relationship/interacting skills.

11.8 Summary

Succession planning is a dynamic tool that can be used to help preserve the future of an ES&H organization. An organization that does not plan for management and professional succession is in danger of losing its competitive edge. Planning to keep an organization equipped with the appropriate skills is never an easy process. Today's businesses are operating with fewer employees who are asked to contribute more in order to keep the business successful, cost-effective, and maximizing company profits. This practice can, and has, added complexity to the succession planning process since there may be little time available to train and mentor successors. Developing and implementing a good succession plan takes time, effort, and commitment from the leadership team. The succession strategy model introduced in this chapter can help guide managers through the process in developing an effective succession strategy.

12

Technology and the ES&H Profession

12.1 Introduction

Today's business environment is for the most part managed and implemented through the use of technology and technological advances. The ES&H profession is no exception when it comes to taking advantage of using technology to improve the way business is conducted. Many managers and professionals were taught how to perform calculations, provide direct field support, understand and recognize regulatory noncompliances, monitor environmental conditions using handheld instruments and manually recording data, and conduct meetings in a room with all meeting participants. In today's world ES&H is routinely conducting business using electronic devices and software that is tailored to provide ready answers for field estimates, and meetings are conducted via telephone or videoconferencing.

The modern business environment is moving toward a paperless society and so should the progressive ES&H program. Data files are electronically stored, transmission of data and communication among divisions and companies electronically occurs, and even performance appraisals and training are tending to be electronically performed and transmitted because of increased efficiencies and ease by which the information can be retrieved and reviewed.

Generally there is a composite of technological knowledge among management and professionals. Professionals that are late in their career may have a tendency to only use technology for word processing and presentation purposes. Younger professionals generally have a greater passion for, and understanding of, the capability and flexibility of the technology applications available to use in the profession, and greater capacity to integrate technology. In particular, for the ES&H manager and professional, technology can be used as an effective marketing tool when interacting with prospective clients. The ability to be able to understand and use various technological devices is being recognized as a valued skill in today's modern work environment. By understanding some of the basic principles of how ES&H tasks can be electronically performed, management and professionals can be more successful and effective when conducting business.

12.2 Use of Technology in the Workplace

There are many uses of technology in the implementation of an ES&H program. Below is a list that contains some of the more common uses of technology. Each of these uses is further expanded in this section; however, the number of uses of technology in the workplace are endless.

- Search engines
- Regulations and standards
- Applications used by various disciplines
- Communications and graphics
- Event response
- Employee observational programs
- Training and certification of ES&H professionals

12.2.1 Search Engines

There are a number of search engines that are available on the Internet that are used as a tool to conduct research and information gathering. Search engines are commonly used by ES&H professionals when conducting business. The use of a search engine for conducting business and research is heavily relied upon today and will be more heavily relied upon in the future. Whether the professional is keeping up with papers recently published on a particular topic or researching a particular contaminant and associated health concerns, the use of search engines has taken the place of professional books and journals. Search engines also provide the capability of interfacing with professionals around the world and are a method to investigate research and standards that are internationally used.

In today's environment the ES&H professional is no longer constrained by the knowledge that is within the United States, but with the use of search engines, knowledge from around the world is available at his or her fingertips. Because of the common use of search engines in society, professionals are expected to have a greater knowledge of the profession than traditionally demonstrated through academic learning.

12.2.2 Regulations and Standards

The use of electronics in learning, in particular with regulations and standards, has become a common practice within the ES&H profession. Whether the professional uses a smartphone, tablet, or laptop, electronic devices have replaced the use of books (e.g., Code of Federal Regulations) because of the flexibility provided by these devices. The ES&H professional can carry his or her smartphone into the field or in a meeting and have the regulations and standards readily available for use. Most professionals have the common regulations bookmarked; however, if the ES&H manager or professional is responsible for more than one discipline, a variety of associated regulations are bookmarked and can be used at any time. Because of the ease by which regulations and standards can be researched, and the small size of these devices, technology has enabled the ES&H professional to be more efficient and productive in the field or wherever he or she may be conducting work.

All regulations associated with the Occupational Safety and Health Administration are electronically available, along with those associated with the National Institute of Occupational Safety Health Administration, American Conference of Government Industrial Hygienists, Nuclear Regulatory Commission, Department of Energy, Department of Defense, and Environmental Protection Agency. Common standards are also available, including those from the American National Standard Institute, Life Safety Code, and National Fire Protection Agency. Others are either free or can be purchased for a minimal cost.

All of these regulations can be electronically downloaded if desired and are readily usable by the professional at any time or place. In addition, in most cases companies can purchase site licenses that allow more than one person to electronically access and use the standards.

12.2.3 Applications Used by the ES&H Discipline

There are a number of electronic applications used by the ES&H professional. There is considerable variability in the actions performed by these applications. Applications are available for performing general calculations, calculations specific to a topical area or problem (such as determining parts per million), or conversion of units. Applications that are commonly used include the following:

- Regulations and standards
- Ventilation calculations
- Heat stress and workload calculations
- Noise exposure determinations
- Radiological contamination and exposure calculations
- Radiological shielding calculations and modeling
- Unit and temperature conversions
- Determination of a regulatory compliant slope
- Self-assessment tools and spreadsheets
- Occupational medical assistance
- Assistance in determining workplace injury and illness recordability and statistics
- Project scheduling tools

There are an infinite number of applications and uses for electronic devices, and more are being generated on a daily basis. It is recommended that specific personnel within the organization be tasked with spending time on a routine basis researching and keeping current on applications that can make the business more efficient and effective.

12.2.4 Communications and Graphics

One of the primary uses of technology is as a communication tool among clients, management, and professional colleagues. The use of computers, tablets, and smartphones has become an invaluable method for communicating everyday activities, emergencies, and just general communication between two or more people. It is not uncommon for personnel in meetings to use tablets for keeping meeting notes, and it is a common practice for meetings to be conducted via telephone or videoconference. Personnel use tablets in the field to record data, and then transfer the data when returning to the office at the end of the day. Many of these methods of communication have reduced errors resulting from manually transferring data and have resulted in significant cost savings to companies.

An emerging trend being used by the safety and health professional is the use of videoconferencing with the company's medical provider. This technique is being used to provide an independent medical evaluation by a doctor of the company's choice and allows greater flexibility in injury and illness case management. It must also be recognized that although electronic means of communication have become a common practice in industry, the ES&H professional needs to be diligent in ensuring that the message being communicated via electronic means is being received in the proper context.

The use of graphics in today's work environment is extremely important because it allows the ES&H professional a more efficient method to communicate ES&H messages to workers, supervisors, managers, regulators, and outside stakeholders. There are a large number of graphics programs available for use with ease, including art that is readily available on the Internet. ES&H messaging can be extremely effective when communicated via a graphical presentation. In addition, because of the variety of electronic programs available for generating graphics and pictures, the ability to create and target the message to a particular audience has improved—the only limitation is the imagination of the originator.

12.2.5 Emergency Response

The ES&H manager is often called to respond to and manage different types of issues and emergencies. Examples of types of issues and emergencies that the ES&H manager can be involved in may include the following:

- Workplace injury or illness
- Vehicle accident
- Fire event
- Weather-related event
- Radiological event
- Chemical event
- Environmental spill
- Area and site general emergencies

Frequently, the ES&H manager or professional is called to control the scene, determine potential personnel exposures and extent of damage, and coordinate/manage response actions. The use of technology in responding to and managing events continues to evolve. Some of the more common devices used during emergency response include the following:

- Videoconferencing and transmitting video from the event scene or with medical personnel
- Digital photography sent via smartphones and tablets
- Portable smart boards and computers
- Cameras and live feed into command centers
- Search engines for researching and developing plans of action
- Dedicated shared media storage areas for direct access and use in case of emergencies

If the action involves some form of communication, it can generally be managed with ease through the use of electronic technology.

12.2.6 Employee Observational Programs

The use of employee observational programs in the identification and mitigation of workplace hazards has increased in popularity over the years. Some of the more successful employee observational programs have incorporated technology into program implementation. Electronic forms are available for employees to use in recording their observations, and this information is documented via a smart tablet or phone. There are a number of electronic applications available on the market that can be used to create forms and organize results. These devices are also used to take pictures of the specific observation so that someone else is able to understand what was observed. Once the employee is finished with conducting the observational routine the data can then be electronically transferred to a share drive, whereby the information can be organized, summarized, and potentially trended.

A key for successful use of technology in employee observational programs is that the electronic device used to record the data must be user-friendly. Most people in today's society are familiar with the basic function of computers and tablets; however, the depth of knowledge varies among users, so any device that may be used by a large number of people must be user-friendly. Some of the applications have accompanying tutorials that assist personnel in their use.

12.2.7 Training and Certification

Training and preparing for professional certifications is routinely conducted using electronic technology. Depending upon the certification being obtained, there are several companies that have produced software and applications based on previous certification test questions that are useful in preparing for the certification exams. It is worthwhile for ES&H managers and professionals to familiarize themselves with the professional society websites, as they are a primary means for obtaining information related to certifications. The profession has replaced hard text and study materials with electronic means of learning, to include exam preparation material, administering of the certification exam, and exam results feedback.

12.3 Types of Technological Devices

Every day it seems there are modifications made to existing technology, or new technological devices are being developed. In particular, this evolution is most pronounced in the use of instrumentation and process monitoring devices. Discussed below are some of the more common devices used by ES&H management and professionals in the industry today.

12.3.1 Smartphones

A smartphone not only allows telephone calls to be made, but also has features that resemble some of the functions performed by a computer. The use of smartphones in business has consistently evolved over the past 10 years. It is common that workers communicate using their smartphones either by texting to another smartphone, e-mail that directly ties to the work electronic mail system, videoconferencing, or search engines and research while in the field.

For the ES&H professional, the use of smartphones enables instant communication and access through texting, use of a camera to record conditions, instant access to electronic files and e-mail for conducting business, and research on chemicals of concern. Because of their size, they are easy to carry and are a convenient method for communicating a small amount of information while still maintaining the information in a digital format.

12.3.2 Laptops and Tablets

Laptop and tablet computers are commonly used in both the office and field environment. Both devices offer flexibility in performing ES&H-related activities, and significant efficiency can be gained in reducing the length of time it takes to perform an activity, such as data gathering and presentations.

Laptop computers generally have larger screens and greater memory that can be used for the manipulation and storage of data. Many of the more common word processing and graphics software programs tend to be more easily installed and used on laptop computers. Laptop computers provide flexibility in providing a traditional computer format and workstation; however, laptop computers are not as easily transportable as tablet computers.

Tablet computers are commonly used in the field for collection and storage of data because of their light weight and small size. Most tablet computers come with a camera, so the operator can take pictures and attach them to a file in the field, and then download the information to an office computer for future use. All tablets come with the ability to connect with Wi-Fi through a wireless network for transmission of data. This allows personnel great flexibility in how the tablet can be used in making their jobs more productive. Tablets are a great way to gather a large amount of data; however, manipulation of the data is not as user-friendly as on a laptop or more traditional desktop computer.

In recent years the computer industry has released laptop computers that can also be used as tablets. Although it is not as common to see these types of devices being used in the business environment, it is envisioned that as more devices can be consolidated while maintaining all the functionality, this will be more cost-effective for companies and their use will gain in popularity.

12.3.3 Portable Instrumentation

Portable instrumentation is a staple among tools used by the ES&H professional in determining field conditions. Whether the ES&H professional is monitoring oxygen levels in a confined space or contamination levels in a radiological zone, the use of portable instruments is invaluable in the decision-making process when protecting personnel and the environment. Every year new instrumentation is unveiled that is smaller in size, can reach a lower level of detection, or can perform more than one function. It is recommended that the ES&H manager and professional attend one trade or professional conference a year to stay current on what new technology is being marketed.

Because of rapid improvements being made in portable instruments it is recommended that prior to purchasing new instrumentation the following questions be asked to better understand needs and equipment usage to ensure the best instrument is procured:

- What are the costs associated with the instrumentation, software, and any routine/anticipated upgrades?
- What are the technological parameters (i.e., temporary memory, downloading capability) that the instrument must meet?
- What interface issues are there with the existing computer equipment/technology?
- Will the instrumentation require the purchase of another computer?
- Does the manufacturer provide technical support associated with operation of the software and networking issues? At what cost?
- Will the company's information technology (IT) group support purchasing and maintenance needs?
- Are there any IT issues that would prohibit the interfacing of the technology to the existing company computer networking system?
- What are the environmental conditions that the equipment will and can be used in?

The questions listed above focus on the issues that are associated with purchasing electronic monitoring instrumentation; however, the ES&H professional should always understand the contaminants to be monitored and environmental conditions in which the instrumentation will be expected to operate.

12.3.4 Simulation Devices and Digital Imaging

The use of simulation devices and digital imaging is gaining momentum for training and operational purposes. There are several devices used within the industry for training that can simulate environmental or operational conditions that could exist in the workplace and demonstrate how personnel should react to various conditions when encountered. Whether the simulation is to improve emergency response actions or everyday work tasks, technology is playing a key role in being able to provide a simulated environment for workers to learn desired safe behaviors. Through the use of digital imaging and various software applications, tools can be created to better enable workers to learn from hazardous conditions. The cost of using these devices and digital imaging capabilities should be evaluated as part of the entire information technology strategy.

12.4 Technology Considerations

Whether you are researching, purchasing, or using technological devices there are a number of considerations that should be explored to ensure efficient operation of the devices. Listed below are a few of those considerations:

- Organization and storage of media
- Validation of data used for reporting
- Security protection of electronic data
- Use of electronic devices in the workplace
- Social media sites and networking within a technological work environment
- Planning cost associated with technological devices
- Overcoming generational differences

12.4.1 Organization and Storage of Media

Data is constantly being generated for use in determining worker exposure to chemicals or radioactivity, regulatory compliance, reporting of performance assurance graphs and charts, training, etc. Depending upon the size of the company and network server capabilities an ES&H organization may create shared electronic storage drives assigned to different functions or project teams to facilitate information sharing.

Shared electronic drives are a great way for more than one person to store and share documents and data. In some companies the cloud concept is being utilized to even further expand their technological memory capabilities. Limitations in the use of a shared drive include making sure one person is assigned as the drive administrator, and that this person routinely manages and evaluates how the drive is being utilized and data stored. Use of a shared drive for personal information, or files from former employees, should be managed and deleted if required. Because of the regulatory aspects of the ES&H profession it is recommended that data stored on any technological device be saved to an additional device, such as secondary memory devices with enough memory to ensure the data is always retrievable. Secondary memory devices include such items as portable jump drives, compact disks (CDs), and portable external hard drives.

12.4.2 Validation of Data Used for Reporting

A common use of data collected by the ES&H professional is to determine whether personnel were exposed to chemical, biological, or radiological hazards or whether an environmental spill has exceeded regulatory requirements for reporting. This data is provided to the individual worker, management, regulatory agencies, medical providers, and in some cases outside stakeholders. The data can also be used in the future as a legal record, and therefore must be subject to independent validation. Should inaccurate information be provided, there are potentially significant liabilities that could exist for the company from a financial perspective, as well as to the company's reputation. When validating data it is important that the person performing the validation activity be independent of the data collection and evaluation process, trained, and demonstrate a proficiency in the methods used to collect the data, and understand the purpose and use of the data.

12.4.3 Security Protection of Electronic Data

Because much of the data generated by ES&H professionals has the potential to be used in the future for liability claims and regulatory compliance purposes, ensuring the data is not tampered with or destroyed is extremely important. The use of passwords and locking of spreadsheets and databases cannot be overemphasized. Not having data to justify a decision made related to environmental compliance, safety and health, and radiation safety can result in a claim or unfavorable judgment imposed against the company. In some cases, the ES&H manager or professional himself or herself can be named in lawsuits, so there is a need to have data to support case management and other potential worker or environmental exposure liabilities. Additional security measures are also a good idea when electronic data files are stored on a shared network drive to avoid data compromise.

12.4.4 Use of Electronic Devices in the Workplace

When purchasing or using electronic devices in the workplace it is always a good idea to check with the information technology group within the company to find out if there are any restrictions with respect to network interfacing. Generally, there are restrictions in how the data may be downloaded and who can access it. In addition, the use of personal electronic devices is often restricted or prohibited in the company setting. Most companies do not allow the use of a personal cell phone in accessing electronic files from the company server.

12.4.5 Social Media Sites and Networking within a Technological Work Environment

One of the primary means of maintaining contact with fellow ES&H professionals is through the use of social media and networking websites. There are a number of sites that most people are familiar with that can be used for keeping in communication with college acquaintances, fellow professionals, national organizations, and professional societies.

All professional societies maintain their own web pages, and that is a primary method for understanding requirements associated with professional certifications. There are also websites for networking among professionals within their own company, within their own profession, and even within their physical location. Networking is important in bringing professionals together to facilitate problem solving, sharing of knowledge and experience, and connections that can lead to professional advancement opportunities.

12.4.6 Planning Costs Associated with Technological Devices

The ES&H manager is responsible for identifying and understanding technology needs associated with performing his or her defined tasks within the company. Because of the rapid rate at which technology changes, it is prudent for the ES&H manager to maintain and update technology on a routine frequency. Below are some rules of thumb that should be considered when conducting financial planning for the organization:

- Upgrade and replacement costs of software and equipment should be a line item in the annual department budget.
- Plan on allocating approximately 5 to 10% (depending on the equipment to be purchased) of the total annual department budget to software and equipment upgrades/replacement.
- Technological training of personnel should be included in the annual budget of approximately 1 to 2%.
- Information technology support should be budgeted at 1 to 2% of the annual budget. If significant network interface support is needed, or a database is needed to be developed, the budget allocated for support should be increased after seeking an estimate for development support.

12.4.7 Overcoming Generational Differences

As part of any work environment, the knowledge and familiarity of ES&H professionals varies. Many late-career professionals today were taught with books and have limited experience with technological devices used in the workplace. The ES&H professionals graduating today are generally more familiar with how to manage and work within a technology-driven environment because many of their classes are taught online and with technology as part of the learning experience.

Because of the differences in learning styles and exposure to technology, the ES&H manager needs to be sensitive and understand that all of the professional staff may not possess a common understanding of the technologies that are being used and be proactive in training personnel who are less technologically advanced.

12.5 Cost–Benefit Analysis

A cost–benefit analysis is used to determine the cost of embarking upon a particular technology against the benefits received by the company should the technology or software be purchased. It is recommended a cost–benefit analysis be performed to better understand the benefits associated with the equipment in order to be able to effectively communicate these benefits to managers and professionals outside the immediate organization. The cost–benefit analysis should be based on both quantitative and qualitative input. Below are examples of both quantitative and qualitative parameters that should be considered:

Quantitative parameters

- Calculated efficiencies gained in productivity and associated man-hours
- Up-front investment costs
- Annual required budget for anticipated software and hardware upgrades
- Timeframe that the equipment will be useful
- Training costs, including those associated with direct staff and those associated with the manufacturer providing the training (if needed)

Qualitative parameters

- Ease of operation and existing knowledge base of equipment operator(s)
- Is the equipment going to be used for regulatory monitoring purposes?
- Cost acceptability by management
- Does the equipment meet more than one need of the company?
- Can the equipment and expertise be used for marketing purposes?

Figure 12.1 is a template that can be used or tailored by the ES&H manager or professional in performing a cost–benefit analysis.

Equipment Evaluation	Cost–Benefit Parameters								
	Calculated efficiencies gained in productivity and associated man-hours	**Upfront investment costs**	**Annual budget for software and hardware upgrades**	**Equipment life expectancy**	**Training costs**	**Ease of operation**	**Intended use of equipment**	**Cost acceptability**	**Use for marketing purposes?**

FIGURE 12.1
Example cost–benefit analysis.

12.6 Summary

The use of technology has had a profound impact over the past 20 years on the ES&H profession. Through the use of the Internet and digital communications ES&H managers and professionals are able to virtually communicate worldwide. The ES&H office used to consist of bookshelves with reference books. Today, the ES&H professional only needs a computerized device and an Internet connection to access a wealth of information, perform work in the field, rapidly respond to an event, and support employees in performing their work in a safe and compliant manner.

As part of managing and conducting work in the field, the ES&H professional needs to understand his or her technology needs and what the vision is of the company in terms of promoting the use of technology in the workplace. As technological innovations continue to be discovered, so too will the effectiveness of the ES&H manager and professional be improved in providing a more cost-effective means to provide a safe workplace.

13

Culture in the ES&H Work Environment

13.1 Introduction

Culture can influence the behavior of members within an organization and affects many aspects of organizational life, such as the distribution of rewards, promotion, and how people are treated. For example, if workers observe through actions of management that production is valued higher than safety, they will begin to shift the values that were originally instilled and adopt the new values being demonstrated by management. Substantial influence can be placed on an organization by its culture because the shared values and beliefs that are present within a culture represent important variables that guide behaviors (Alston, 2013).

Many companies have a mission statement and core values that form the fundamental basis of how the company is managed and operated. The mission statement and core values should directly correlate to whether a company is proactive in managing its culture. Management actions and behaviors should be influenced by the mission statement and core values. The success of business is often measured by the amount of profit and growth sustained over 1, 5, or even 10 years. In some companies, the means to achieve business success may or may not be consistent with the mission and core values of the company.

As illustrated in Figure 13.1, when there is a perception (demonstrated through daily communications, actions, policies, and procedures) that the methods used to achieve success are inconsistent with the mission and core values, then that perception influences and drives the behaviors of management and the workforce. In particular, the culture of an organization and company directly influences whether core values are being appropriately executed. For the ES&H manager, the impact of organizational culture when developing and executing various programs is significant and is an influential element in how business is transacted.

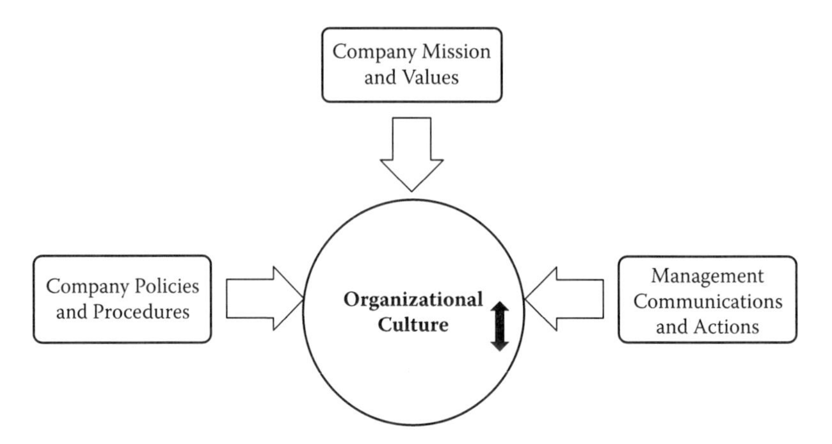

FIGURE 13.1
Culture influencers.

13.2 Impact of Culture in the ES&H Environment

Culture influences each area and discipline managed by the ES&H manager. Whether implementing a rewards and recognition program or addressing safety concerns in the workplace, the influence that culture plays in success of the ES&H environment can make the difference between whether a program will be successful and beneficial, or one that does not achieve meeting the needs of the company and protecting the workers and environment. Often the impact of culture of the various ES&H disciplines may first be observed by either the improvement or decline of performance metrics and incidents that occur in the field. The impact of culture on each of the ES&H disciplines is discussed below.

13.2.1 Safety and Health

The basic fundamental premise of the safety and health discipline is to protect the workers from industrial and chemical hazards in the workplace. Also included in this function is occupational medicine to ensure the workforce is healthy and qualified to perform their jobs. An initial indicator of how well the culture of the organization or company is is often observed in the total recordable case (TRC) rate and days away case indicators.

The approach and attitude of the employees in embracing the organizational culture is often linked to their approach to hazard identification and mitigation. In particular, it is often found that as the TRC increases there is an observed degradation in the organizational culture. If the organization culture is not healthy and workers do not observe that fundamental organizational beliefs and values are being demonstrated by management and supervision, the workers themselves will start to disregard and not believe in the organizational culture. In the case of safety and health, workers may not raise concerns when warranted, may not promote safe work practices, and employee involvement may decrease and workers may become more frequently injured.

A startling statistic is that in 2012, 4,628 workers were killed on the job (Occupational Safety and Health Administration (OSHA) website). All of these deaths could have been prevented, and it can be assumed that the safety culture of many companies contributes to death and workplace injuries every year. The OSHA website is an excellent resource for additional information regarding the link between workplace safety and safety culture.

13.2.2 Radiological Protection

Radiological protection strives to reduce the amount and frequency of exposure of employees to man-made or naturally occurring radiation. The influence of culture in implementing a successful radiation protection program is significant since the impact of not having a successful program can harm not only the worker but also the environment and the public. The influence of culture on the nuclear industry has been well documented. One such example is the Davis–Bessie nuclear incident.

On February 16, 2002, workers at the Davis–Bessie Nuclear Power Station began a standard refueling procedure that called for the shutdown of the plant and an inspection of the reactor itself. During the course of the inspection workers were examining the nuclear reactor vessel head and discovered a football-sized hole. Borated water within the vessel had leaked from its container and eaten away nearly all of the 6½ inches of steel over the football-sized area, and the possibility of a reactor rupture was dangerously likely if the plant had been allowed to go back online.

After the discovery of the hole it was determined that the damage had occurred over the course of 6 years, meaning that inspections in 1998 and 2000 had somehow managed to overlook the dangerous eroding of the steel head. The Davis–Bessie reactor was shut down from March 2002 until 2004 to undergo further inspections and repairs. Over the course of the inspections a multitude of other design flaws were discovered.

On January 20, 2006, FirstEnergy agreed to pay fines of $23.7 million for violation of a variety of safety codes. In addition, two past employees and a former contractor were formally charged. There is a recognized belief that a degraded safety culture was a contributor to causing the Davis–Bessie event. Over the past decade the Nuclear Regulatory Commission and International Atomic Energy Agency have established well-defined approaches and methods for managing culture in the workplace.

13.2.3 Environmental Protection

The function of the environmental protection functional area is to protect workers and the environment from harmful chemical spills and releases. It is a broad subject because included within the discipline is the management of chemicals under the Superfund Amendments and Reauthorization Act (SARA) Title III and the Emergency Planning and Community Right-to-Know Act (EPCRA). The manner by which some chemicals are stored, used, and released is strictly regulated under the Environmental Protection Agency (EPA).

The influence of culture in the environmental protection discipline is profound because the values and beliefs that are embraced by the organization directly influence the attitudes and actions of workers managing chemicals and adhering to regulatory requirements. Under most federally regulated programs, companies and, in some cases, employees can be held criminally liable for chemical releases, spills, and mismanagement of chemicals. The classic example of the impact of culture on environmental protection is the Bhopal incident in India in 1984.

The Union Carbide facility in Bhopal opened in 1969 and produced the pesticide carbaryl, which was marketed as Sevin. Ten years later the plant began manufacturing methyl isocyanate (MIC), a cheaper but more toxic substance used in the making of pesticides. It was the MIC gas that was released when water leaked into one of the storage tanks on December 2, 1984. The first effects were felt almost immediately in the vicinity of the plant.

More than 3,800 people, both workers and the ensuing community, were killed as a result of the accident. There were many causes of why the incident occurred, but the culture of the plant was identified as a contributor to the accident.

13.3 Methods to Evaluate Culture

The overall culture of a company, organization, or group can be evaluated using a number of methods. These methods may be either directly or indirectly a measurement of culture. As illustrated in Figure 13.2, below are some of the more frequently used methods for evaluating and measuring culture in the workplace. Each of these methods is further discussed below. Often more than one method is used to gauge the health of an organizational culture:

- One-on-one daily discussions
- Interviews
- Surveys
- Performance indicators

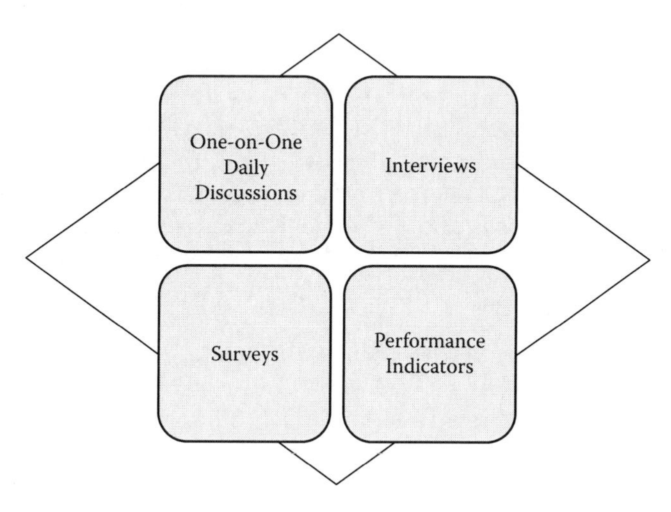

FIGURE 13.2
Methods to evaluate culture.

13.3.1 One-on-One Daily Discussions

In the age of electronics, daily verbal communication between workers and management, including professionals, is an extremely effective method for evaluating culture and understanding the perceptions of workers of the company's values and beliefs. In many companies verbal interactions in the workplace are decreasing in frequency and communication is progressing toward e-mail and texting.

The electronic means to communicate has gained momentum because of the ease by which communications can be accomplished. However, if used as a primary means of communication, it may not facilitate the feedback-response loop from the receiver. Over time, the worker can perceive the communications to be impersonal because they do not allow the receiver to communicate his or her feedback.

Unlike electronic means of communicating, one-on-one daily discussions yield a wealth of information to the ES&H manager. Feedback from the workers can provide an indication of whether they concur that the hazards that have been identified and the types of mitigation techniques that are used to protect them are adequate.

Perceptions that employees may have regarding the company and management team can directly be gleaned from the worker during discussions, along with suggestions for improving the workplace culture. Using the feedback received from the workers is a great way to improve culture and gain buy-in when changes are made. Daily one-on-one interactions are one of the most effective means to gauge the culture of the company or organization.

13.3.2 Interviews

Interviews have a significant role to play in the assessment of culture. They are commonly used as a means of providing data that will assist in a survey design or to qualitatively explore the issues identified as a result of a written survey. An advantage of the interview process is that the interviewee can use his or her own words and expressions. It also allows for a greater flexibility in questions, with the possibility for follow-up questions, making it easier to get to the deeper meanings of statements and to clarify ambiguities in responses.

A difficulty with interviews is that the information and process have some variability, thereby making it more difficult to perform a direct comparison. Interviews are also relatively time-consuming, usually based on only a limited sample. This can make it difficult to generalize results for the whole organization. Interviews can provide more in-depth detail than normally gained using a survey.

Interviewing groups of people or individual interviews of workers can provide great insight into the health of the organizational culture. Interviews can be conducted in one of two ways:

- Individual interviews
- Group interviews

13.3.2.1 Individual Interviews

Individual interviews provide a forum whereby the interviewee feels more comfortable discussing issues more personal to him or her. Interviewees have a tendency to speak more freely if other people are not in close proximity and they feel that what they communicate will be kept confidential. Interviewing workers individually is a great technique if investigating whether the culture issues are unique to a working group rather than systemic within the organization.

13.3.2.2 Group Interviews

Group interviews provide a means by which to reach a larger group of workers in a limited amount of time. The group interviews can be structured by various means: large group setting or individual focus groups that share common work activities or management personnel. A focus group is the meeting of a small group of workers (four to eight) guided by a moderator. The strength of the focus group interview is also its weakness.

In conducting the interviews the interviewer relinquishes some control over its form, content, and development—it is a complete iterative process. A group interview, unlike the individual interview, is inevitably a less anonymous process. Participants may be concerned about expressing critical views in a group setting, driving them to be unable or unwilling to open up and provide critical information.

13.3.2.3 General Considerations When Interviewing

The frequency and depth of interview will vary depending upon the intent and use of interview results. When conducting interviews it is important to consider the following key points:

- Conduct the interview in a worker-friendly environment.
- Introduce yourself, state the purpose of the interview, and ask at the beginning and end of the interview if the interviewee has any questions.
- Humor is always a good way to start the interview or any other conversation to put the interviewee at ease.
- When conducting the interview phrase the questions in nonleading, open-ended sentences.
- Use the same question bank for each interview conducted.
- Summarize the results; identify common themes.
- Share the interview results with the interviewees.

13.3.3 Surveys

Surveys can be a useful method to assess culture. The extent of such surveys will vary depending on the size and organizational structure of the company. Survey results can indicate employee beliefs, attitudes, and satisfaction with key attributes and suggest ways to strengthen the culture. Presurvey communications can be a very important aspect of such a method for evaluating culture (Energy Facility Contractor Group, 2009).

A written survey can provide a number of benefits:

- The information can be quantified and results compared between groups over time.
- They are relatively easy to administer, minimizing work disruption and encouraging a high response rate.
- They provide clear data, which can be rapidly analyzed.
- Respondents can remain anonymous, encouraging them to express critical views without fear of adverse consequences.
- The provide a precise and reliable reflection of the total population.

Potential drawbacks to the written survey include the following:

- Answers to questions may not reveal the full depth of unconscious assumptions that underpin beliefs, values, and attitudes.
- Poorly worded questions can produce misunderstanding, or inadvertently prompt the more socially acceptable answer.
- With questions generally limited to certain categories, it is also difficult to obtain information about the various aspects of a situation. That can make ambiguities difficult to deal with.
- Authoritative statistics can be misleading.

If interested in performing a survey-based evaluation of culture there are commercially available electronic means by which to administer the survey (e.g., Survey Monkey®).

13.3.4 Performance Indicators

Performance indicators can provide continuous feedback on the health of the organizational culture as opposed to assessments, which represent snapshots of the culture at the time the assessment is conducted. Care needs to be taken in selecting indicators to avoid overemphasizing any particular aspect. A broad range of performance indicators may be needed because of the complexity of the organizational culture that may be present.

Some potential advantages and disadvantages of using indicators to measure organizational culture are shown below:

Advantages

- The use of indicators allows trends to be detected.
- Managers pay greater attention to what is being measured, and the use of culture indicators will increase their interest in the concept.
- Culture is an important aspect of the organization; hence, it should be treated like other important business aspects and measured. It should be identified as one of the core values within the company.

Disadvantages

- Organizational culture can be complex and no direct method to measure it exists; thus, any attempt at measurement should be indirect.
- Some elements of organizational culture, such as basic belief, may be unconsciously held and present great difficulties for measurement.

- Managers may feel that they have little influence over trends in organizational culture when they have limited understanding of the concept of culture and the period of time required to effect a change.

13.4 Culture Improvement Initiatives within ES&H

As an ES&H manager there are number of initiatives that can be launched to improve organizational culture both within the department and within the company. Generally, a degradation of the organizational culture is most noticeable by the ES&H management team because of the impact a poor organizational culture can have on ES&H performance indicators. Some initiatives that can be useful to the ES&H manager and professional in improving organizational culture are highlighted and further discussed below:

- Clearly defined core values and company policies that establish the expected behaviors of management and the workforce and communicate the importance and influence of organizational culture on the work environment
- Development of key performance indicators that are used to measure organizational culture in the workplace
- Leadership/supervisor training
- Employee involvement committees
- Communications plan
- Focused campaigns

13.4.1 Defined Culture Values

The fundamental principles for how a company will operate, including influences on management decision making, are based on established company core values. For example, a core value may be open communications. How that core value is implemented and perceived within the company directly influences organizational culture in the work environment. By establishing a core value and subsequent policies and procedures that are reflective of a robust organizational culture, behaviors and decision making will be reflected in daily work activities.

13.4.2 Key Performance Indicators

Because the health of organizational culture attributes is reflected in individual decision making and behaviors that are qualitative in nature, it can be difficult to directly measure. Therefore, a key performance indicator of culture would include measurable metrics that are collectively evaluated to determine, from a qualitative perspective, whether the culture or health of the company is degrading or improving. Example metrics that can collectively be evaluated in determining whether the company culture is progressing in the desired direction include the following:

- Number of management observations and time in the field
- Turnover rate of personnel
- Number of employee concerns or issues formally raised
- Number of times workers in the field have been involved in the work planning process
- Number of issues identified in a formal corrective action management system

The examples listed above only represent a small fraction of metrics that could be evaluated to provide feedback to the leaders, managers, and workers on the overall culture of the organization. The key performance indicator should be directly tied to the organizational culture core value and tailored to the mission and services provided by the company.

13.4.3 Leadership and Supervisor Training

Interactions of management and supervisors can contribute significantly to the type of culture present in the work environment. Managers and supervisors are considered the face of the organization and ultimately drive how the company will be managed and whether an organization has a healthy culture and work environment. Not only do their actions and communications drive profit and company priorities, but workers look to them for determining what is acceptable versus nonacceptable behavior, standards of conduct, and what is considered the "norm" of the work environment. Consequently, the actions and behaviors of the leadership team are a key element of a healthy organizational culture.

In many instances personnel are promoted into positions of management and supervision with no additional training other than that received as part of their craft or discipline. Often leaders of a company assume if a person performs well technically, he or she can successfully manage people and the work that is being performed. This is frequently a flawed assumption because the newly promoted manager may not have the soft skills necessary to successfully manage the organization or group.

An important element to improving the organizational culture is to provide leadership training to managers and supervisors. The training should build upon the mission and core values of the company and teach leadership attributes and skills for managing conflict. Below are attributes that promote a healthy organizational culture:

- Management evaluates and provides feedback on performance and responsibilities and reinforces expectations to ensure desired behaviors are being displayed.
- Personnel are held accountable.
- Management recognizes and rewards positive behaviors.
- Shortfalls in meeting requirements are promptly reported.
- Management communicates with their employees. Communication is on a personal level and management spends time with employees in their work environment.
- Supervisors and managers are out in the field on a frequent, if not daily, basis.
- Managers and supervisors set an example of desired behaviors. They "walk the walk" and "talk the talk."
- Management decisions are conservatively made and based on concrete data.
- Managers include employees in the decision-making process. By including employees in key operational decisions, management is able to gain buy-in, resulting in reduced operational risk.
- Management promptly reports errors, thereby demonstrating the desired behavior of self-reporting.
- Workers promptly report errors because they feel they are in a safe environment that values self-reporting.
- More than one method is available for employees to report issues.
- Managers and supervisors understand and accept their responsibilities inherent in mission and company accomplishment. Supervisors do not depend on supporting organizations to build a healthy organizational culture into work activities.
- Managers and supervisors have a clear understanding of the company core values and performance objectives, and this is demonstrated in their daily behavior (e.g., respect, integrity, honesty, fostering trust).
- Managers and supervisors support and promote continuous learning.
- Management has a succession planning strategy to sustain a technically competent workforce.

13.4.4 Communications Plan

Once organizational culture issues are identified it is imperative that the company be open and transparent about the issues and path forward. Open and honest communication is a key element in improving the culture and building trust in the leadership and management team. Targeted and focused communication is the most effective manner to identify the specific issues and path forward. Elements to consider when developing an effective communication plan include the following:

- The intended audience for the message.
- Specific issues to be addressed along with defined corrective actions. The corrective actions should have defined completion dates.
- Types of messaging approaches to be used. These include Internet forums such as shared websites, blogging, and other shared electronic resource mechanisms, conference room posters, banners, messaging on electronic or hard copy reader boards, or company publications. Another example is use of a "roving reporter" that gathers feedback from employees, and the information is videotaped for future media campaigns.
- Methods to be used to provide routine status to employees of progress being made to improve the organizational culture.
- Central location for communicating the status of the improvement initiative.
- A defined celebration of meeting an improvement goal (as appropriate).

The content of the communications plan should be tailored and focused on communicating improvements that will enhance and promote a stronger organizational culture.

13.4.5 Focused Campaigns

Focused campaigns are initiatives that include a series of messages or an exercise, conducted over time, to highlight an area needing attention. Examples of focused campaigns include the following:

- A performance metric initiative that focuses on increasing the amount of reporting worker feedback or participation in the work planning and control process
- An initiative by management to increase the amount of time spent in the field
- An initiative to improve worker participation in a worker safety or environmental observation program

These are just a few examples of various campaigns that can not only improve the organizational culture, but also improve performance associated with an area of concern from an ES&H perspective.

13.4.6 Employee Involvement Committees

The use of employee involvement committees as an initiative to improve organizational culture is one of the most powerful tools available to the leadership team. The ES&H manager often develops close relationships with the workforce due to the nature of his or her job since he or she is viewed as having some responsibility for an employee's health and well-being.

Generally there is an element of trust established between the ES&H professionals and the workforce. Consequently, the ES&H manager is often readily accepted and trusted by the employee committees and can be useful to the company when trying to build a cohesive working team and improving organizational culture.

Employee involvement committees are a great forum for brainstorming and the generation of ideas for initiatives that may have not been considered by management. Because of the diversity of the workforce and the committee they are able to identify initiatives that are directly relevant to their daily work environment and tasks. Often the employee committees develop initiatives that are more readily accepted by other employees. When the initiatives are implemented, it demonstrates to the employees that management values and trusts their opinion.

The number of employee involvement committees will vary by the size of the company, but there is generally always a safety and health committee and an employee recognition committee. Both of these committees are excellent forums for soliciting assistance in addressing one or more organizational culture issues. As mentioned above, the use of the employee involvement committees should be highlighted in the communications plan and promoted when launching cultural actions.

13.5 Tools for Improving Culture

There are a number of tools available for improving organizational culture. In particular, the book entitled *Culture and Trust in Technology-Driven Organizations* (Alston, 2013) lists a number of tools for use by managers in evaluating trust and culture in the workplace. Listed below are two examples and a brief synopsis from the reference book. A full discussion on these tools, and others, is available in the book referenced above.

13.5.1 Organizational Culture Questionnaire

The purpose of the organizational culture questionnaire (OCQ) is to gauge the type of culture serving as the foundation of the organization's systems. In completing the OCQ you will be providing a picture of important attributes of the organization. The OCQ consists of 28 questions designed to provide an indication of cultural elements applicable to your organization. Figure 13.3 is a reproduction of the survey template. Survey instructions include the following:

Organizational Culture Questionnaire							
For each question given below, circle the number that best describes your opinion to the questions listed.							
1	2	3	4	5	6	7	8
Always	Mostly	Frequently	Usually	Sometimes	Infrequently	Seldom	Never
#	Question						
1	My organization has a clear vision.						
2	The values of my organization are shared by its members.						
3	I believe that my management values my opinion.						
4	Communication in my organization is fluent and flows in all directions.						
5	The managers in my organization recognize and celebrate the success of its members.						
6	The mission and values of my organization are posted for employees to view.						
7	In my organization management celebrates the successes of employees at every level.						
8	The management team is trusted and respected by employees at every level.						
9	Management is responsive to suggestions from employees.						
10	Conflicts are handled openly and fairly.						
11	Employees are motivated to perform their jobs.						
12	Employees understand their job duties and their roles within the organization.						
13	Downward communication is accurate.						
14	The organization's goals and objectives are clear to employees throughout the organization.						
15	Roles and responsibilities within the organization are clear and understood.						
16	My input is valued by my peers.						
17	Employees have the right training and skills to perform their jobs.						
18	Knowledge and information sharing is a common practice for members of my organization.						
19	Disagreements are addressed promptly when they occur.						
20	Morale is high across my organization.						
21	Employees enjoy coming to work.						
22	I feel that I am valued as a part of my team.						
23	Employees speak highly of my organization.						
24	Roles and responsibilities in my organization are clearly defined and understood.						
25	Everyone takes responsibilities for their actions.						
26	My supervisor is a positive role model.						

FIGURE 13.3
Organizational culture questionnaire.

- Read each question carefully and rank your response based on your perception of organizational performance from 1 to 8.
- Fill in the response that you believe most closely represents your organization's performance for each question. Please only select one response for each question.
- Once all questions have been completed use the culture continuum (Alston, 2013) to determine where your culture is located based on the average results continued from the response to the question.

13.5.2 Culture Improvement Plan

In many instances the ES&H manager and the human resource team are delegated the responsibility for developing a path forward for resolving culture issues. One method for defining and implementing corrective actions is through a culture improvement plan (Alston, 2013).

The improvement plan should address areas that were determined to be below the desired outcome, as well as ways to maintain or improve the areas that were identified as strengths. The improvement should provide a detailed road map on how the program, process, and behaviors that are in need of improvement should be addressed. Also included in the plan are measures that will be taken to avoid eroding the areas that were identified as strengths. An example outline for a culture improvement plan is presented below. Please note that the improvement should be tailored to the size of the company, client expectations, and extent of issues.

Example Culture Improvement Plan

X.1 Introduction
X.2 Purpose and Scope
X.3 Areas of Identified Strengths and Weaknesses
X.4 Identified Corrective Actions for Resolving Issues
X.5 Identified Actions to Continue Areas of Cultural Strength
X.6 Funding Required to Execute Recommended Path Forward (optional—could be incorporated into Sections X.4 and X.5)
X.7 Schedule for Completion of Proactive and Corrective Actions
X.8 Follow-Up Evaluation of Success of Proactive and Corrective Actions, Including Criteria Used for Evaluation

13.6 Summary

Management of safety culture is one of the most rewarding and challenging areas for company success that the ES&H management will encounter. It is based on personnel management rather than technical approaches, which can be challenging to a function that is primarily process and technology driven. The impact of a company's vision and mission is directly relevant and an important piece of organizational culture. Because of the diversity of the various functions of the company, including ES&H, organizational culture issues can have a direct impact on the company's scope, cost, and schedule that is directly linked to the productivity and profit of the company.

There are a number of methods available to evaluate organizational culture. Listed above are a few examples, but the bottom line is that the company needs to decide what method is most compatible with its mission and vision.

The number of initiatives available for improving organizational culture are endless; however, it has been well demonstrated that whatever approach is used, employee involvement and ownership of these initiatives are imperative to success. Without acceptance by the employees of these initiatives, they will be less effective than envisioned. Two tools useful in improving organizational culture were identified, including organizational culture surveys and a culture improvement plan.

Although there are many challenges faced by the ES&H manager when facilitating cultural issues, the reward for successfully resolving culture-related issues will result in multiple benefits, including increased employee morale, reduced injury and time away from work associated with on-the-job injuries, increased production, improved customer satisfaction, and most important, a team that is built on sound principles that are focused on the same mission and vision of company success.

14

The Impact of Trust in an ES&H Organization

14.1 Introduction

More and more leaders are recognizing the importance of trust when it comes to being able to conduct business with ease. An organization built upon trust is one of the most important elements in having a thriving business in today's economy. In recent years many influential business executives abused their positions in ways that have managed to erode the trust of workers. Trust has been linked to a variety of positive organizational characteristics and behaviors, such as increased performance and productivity, higher level of cooperation among team members, and increased morale among employees. Many leaders have come to realize that productivity and the ability to gain access to critical information and skills can suffer greatly if the appropriate level of trust is not present within the organization. When workers trust their management team they are willing to go the extra mile to ensure the success of the business.

Many organizations are struggling today because their workers lack trust in their management team and the organization. This lack of trust can have serious negative impact on the business and the bottom line. In many cases, the leaders of those organizations are not aware that trust has eroded or has not been fully developed. Additionally, they may not be aware that trust is not an automatic organizational trait; rather, it must be gained and constantly nurtured. Therefore, they are not focused on developing a trusting culture. Workers don't trust management because of the position or status within the company, as some may believe. Trust is essential in developing mutually dependent relationships that are based upon repetitive actions that yield constant results. It is important for the ES&H staff to be viewed as trustworthy by management and employees in order to accomplish the scope of their task with ease.

14.2 Trust the Building Block for Organizational Success

An organization must develop trust with its employees and customers in order to grow and survive in a fast-paced business environment. Managers must take the initiative to build trust on a daily basis with employees and their customers. Trustworthiness is the foundation of any credible organization, and it begins with the leadership team. Hence, trust building must be an integral part of every organization business strategy. In order to build and maintain trust in organizations a deliberate plan should be developed and consistently implemented. This plan will yield success only when it is included as an integral part of the overall culture-building strategy. The success of organizations may be tied to management's ability to gain and retain trust of the workforce. In order for an ES&H organization to successfully provide the needed support to its customer, a relationship built on trust must be developed, nurtured, and maintained. When an individual trusts another person or organization, he or she is able to be influenced by that individual or organization. Influencing people, programs, and processes is at the core of the task faced by an ES&H professional. It is difficult for the ES&H professional to impact safety and health of the worker when trust is not at the core of the relationship and the decisions that are being made.

14.3 The Impact of Mistrust in an Organization

To complete the discussion on trust we must talk about the effects of mistrust on an organization. A culture where mistrust flourishes is viewed as a low-productivity culture, one where people are focused on "covering their decisions and actions" and have little, if any, desire to work outside of their comfort zone for fear of making mistakes and the associated consequences. An organization where mistrust flourishes tends to be attributed to the caliber of the leadership team, the policies and practices that are in place, and the way those policies are implemented. The most frequent reasons why workers do not trust their management team and organizations include the following:

- Display of abrasive behavior.
- All employees are not treated equally. Favor is shown for some employees in words and action.
- Management's unwillingness to admit mistakes when mistakes are made. When mistakes are clearly seen by workers and the manager fails to admit mistakes, respect for the manager and the company suffers.

- Lack of trustworthy behavior by management. The behavior of the leadership team is inconsistent and extremely employee dependent.
- Lack of accountability. For example, the management team devised and implemented a policy that is not enforced. Employees are not held accountable and no actions are taken in such case where the policy is not adhered to.
- Managers have checked out and are biding time. These managers are not engaged in the business and simply show up for work.
- Hidden agendas. Management has self-serving reasons in mind that are noticed by the worker.
- Self-absorbed. Decisions are made to promote the manager's agenda and not for the good of the organization.
- Lack of employee involvement in decision making.
- Unfair promotion and compensation practices. The policies and practices used for promotions and rewards are not fairly distributed and consistently implemented.

The most frequent triggers of mistrust when it comes to accepting the advice and regulatory direction from an ES&H organization and its members are

1. Lack of demonstrated knowledge in ES&H
2. Inconsistencies in decision making and implementation of program and policies
3. Providing conflicting and unclear instructions when it comes to regulatory compliance
4. An unwillingness to demonstrate flexibility in methods used to ensure compliance when feasible
5. An unwillingness to function as an integral part of the team, resulting in being viewed as "regulatory cops"
6. Functioning like a consultant as opposed to a member of the team seeking to help with finding viable solutions

14.4 The Role of Trust in an ES&H Organization

In order for the ES&H team to be viewed as trustworthy, they must demonstrate credibility in actions and decisions. An organization that has demonstrated credibility has set the stage to gain the trust of others and perform this task with ease. Too often ES&H professionals don't take the time to consider the impact of their decisions and how the delivery of the regulatory compliance message can be negatively perceived by the worker. Recognizing that regulatory compliance is not an option and delivery of that associated message must occur, the ES&H staff should be skilled in communication. When the message is not delivered with clarity and with the appropriate level of facts, it can lead to mistrust on the part of the customer and the workers.

Scenario

The safety and health manager is constantly dealing with workers questioning the decisions and advice given by several of his subject matter experts (SMEs) with regard to safely performing work and implementation of regulatory requirements. The manager is trying to recover from two recent decisions made by two of the SMEs. The first incident involved a misinterpretation of an Occupational Safety and Health Administration (OSHA) standard that resulted in an incident that had the potential to have been catastrophic. An investigation of the event revealed that the SME assigned to make and communicate decisions regarding safety and interpreting regulatory requirements had not completed the necessary training and did not possess the knowledge required to function as an SME. The second incident involved a mistake made by an SME during setup of an air sampling collection protocol to collect personal samples while employees were conducting pipe welding. The SME developed the protocol that called for a particular type of sampling media to be used to collect the samples. Three employees were monitored during the task wearing the personal sampling pump with the prescribed sampling media. At the end of the day the SME collected the sampling pumps, removed the sampling media, completed the appropriate documentation, and delivered the samples to the laboratory. Upon reaching the laboratory and during relinquishing the samples to the laboratory technician it was noticed that the wrong sampling media was used to collect the sample. Therefore, the samples had to be voided. On the next day, the SME explained to the workers that the samples had to be collected again on that day because a mistake had been made in selecting the incorrect sampling protocol and media on the previous day.

QUESTIONS TO PONDER

1. What impact did that decision made by the SME have on the credibility on the SME and the organization?
2. Who was responsible to ensure that the SME had the appropriate training, knowledge, and skills to perform in the role as an SME?
3. How did the use of the wrong sampling media impact trust between the employees and the SME?
4. What impact, if any, did the actions of the SMEs have on the organization?

5. What impact, if any, did the actions of the SMEs have on the safety and health manager?
6. How has the credibility of the organization been impacted?
7. What can the manager do to recover from the impacts of the mistakes that were made?

To avoid the issues brought out in the scenario, ES&H professionals must possess the appropriate knowledge and skills to successfully perform their roles, and be viewed as credible and a part of the overall success of the company or project. The scenario also demonstrates the importance of having a process that will ensure that qualified and skilled professionals are selected to fill ES&H positions.

14.5 How to Establish Trust with Customers

Establishing a relationship of trust with the customer is not optional; it's a must. It takes a committed and trustworthy management team setting the stage and demonstrating the values of the organization to build trust with the customer. Establishing trusting relationships, although it takes time, can be achieved through consistency in actions. It is extremely valuable to invest the time needed to establish a relationship built on trust with the regulators, workers, and customers. The following actions can serve as a mechanism to build trust with customers:

1. Always be open and honest with the customers.
2. Take the time to introduce yourself and establish a relationship with the regulators.
3. Conduct business proactively.
4. Consider the needs of the client when making decisions.
5. Frequently meet with the customer and get feedback.

It is important for the ES&H manager to listen to what other people such as the worker and management team in other departments have to say and try to appreciate and understand their viewpoints on how work can be implemented safely and while remaining in compliance with the applicable regulations, standards, or policies. In order for an ES&H manager or professional to be viewed as credible, he or she must possess and demonstrate sufficient skills. These skills form the basis for the development of trusting relationships that will be necessary and important when conducting business. This relationship is shown in Figure 14.1.

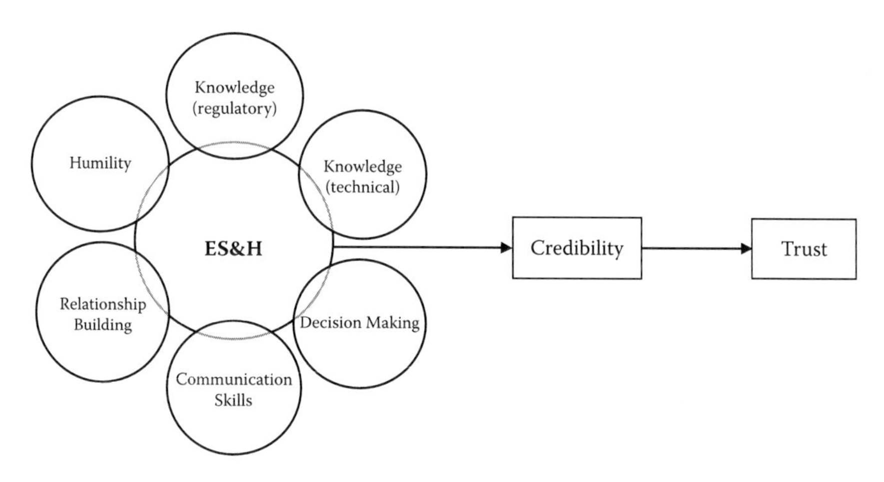

FIGURE 14.1
ES&H trust paradigm.

The characteristics listed in Figure 14.1 can be useful in aiding the staff in connecting with the customer and ultimately gaining and maintaining trust. The trust attributes commonly recognized as necessary factors to consider when designing a culture of trust have been further expanded into a trust-sustaining model that can further assist in developing and maintaining the level of trust necessary to cultivate the culture. These attributes of that model are discussed below and the relationship is shown in Figure 14.2.

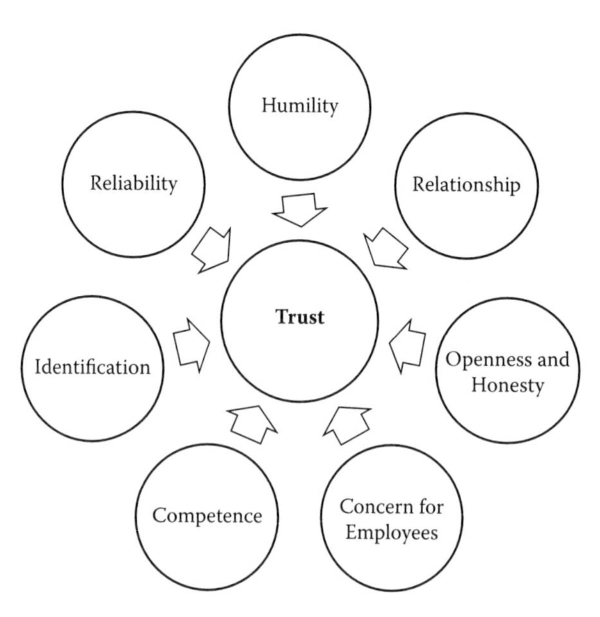

FIGURE 14.2
Trust-sustaining model.

14.5.1 Humility

Humility is a very powerful tool for people who are responsible for engaging others and getting them to follow their lead. It is well established that leaders that lead with humility are more effective than leaders who do not. In an ES&H organization it is necessary not only for managers to have leadership responsibilities, but also each professional and SME since they provide guidance for people to follow. Humility is an essential behavioral characteristic that can be used to enhance one's ability to sharpen his or her skills of persuasion. Skills of persuasion are essential in being able to influence others to follow and buy into a vision.

14.5.2 Relationship

An important aspect of building trust includes the ability to build lasting and effective relationships. People are different when it comes to their need for close personal relations with others and tend to proceed with caution when developing relationships in the workplace. An organization's culture based on cohesive trusting relationships and respect can foster attachment among individuals. Trusting relationships generally provide motivation and bring out the best in people. In the workplace, at times the decision to follow or trust another is based upon preestablished relationships.

14.5.3 Openness and Honesty

Openness and honesty are two important characteristics of a successful leader and leadership team. These characteristics can be demonstrated when commitments and promises are consistently met. To demonstrate openness and honesty, information must be disclosed in totality and delivered with accuracy and sincerity. If the ES&H team is not viewed as being open and honest in communication, their messages may not be accepted and may not be followed.

14.5.4 Concern for Employees

Being sensitive to and understanding people's needs and interests is an important element for building trust. Demonstrating concern for employees deals with the feeling of caring, tolerance, and safety demonstrated through interactions. Safety of the environment and the workers is the primary objective of an ES&H program. It is important for the ES&H professionals and the management team to listen to what other people have to say and try to appreciate and understand their viewpoints. This act by the ES&H team can demonstrate respect for individuals and their ideas.

14.5.5 Competence

Competence refers to qualities such as influence, ability, impact, knowledge, and the ability to accomplish what is needed. It does not specifically refer to the leaders' skills and abilities in the technical aspects of the business. However, when referring to the ES&H professionals and their ability to function in their roles, competence does refer to the technical aspects of their skills. ES&H professionals are expected to display expertise in the area of regulatory compliance and the technical aspects of their jobs.

14.5.6 Identification

The identification attribute refers to the extent to which groups or organizations hold common goals, norms, values, and beliefs associated with the organization's culture. Identification fosters commitment by shaping expectations about behaviors and intentions and leads to certain actions that will support the vision of an organization. Passion results from identification. Without identification there is no passion—and very little, if any, commitment. A higher level of identification results in a higher level of commitment, loyalty, and performance. When the ES&H team identifies with the employees they have been hired to provide guidance and support to, it becomes easier to obtain acceptance and support for a safe working culture and it is more easily embraced.

14.5.7 Reliability

It is necessary that the ES&H professionals are viewed by their customers as being reliable in decision making and the ability to provide appropriate support. Reliability in making appropriate decisions and providing support is key to being able to influence the safety and health culture of an organization. Predictability is also a key component of reliability since it takes the guesswork out of how a person will act or react. Reliability addresses whether or not an individual will act consistently and dependably in communications and interactions. For the ES&H team it is paramount that the customer believes that the appropriate level of support is available when needed.

14.6 The Role of the ES&H Leadership Team in Building Trust

One of the major responsibilities entrusted to the senior leadership team is to set policies and procedures used to determine how business is conducted. Designing a credible ES&H organization begins with the leadership team. There are many benefits realized when stakeholders, customers, and workers trust the ES&H organization that has been chartered to provide safety and health services. These benefits include the following:

- Willingness of employees to follow safety and health policies developed by the organization
- Ease of changes in policies and procedures when there is a change in regulatory standards that may impact facility operations
- Willingness of employees and stakeholders to seek advice on issues regarding safety and health

It is not feasible for a leader to be viewed as a great leader if he or she is unable to gain the trust of others. The same holds true for the ES&H leadership team. They must not only gain the trust of the managers that are their peers in other departments, but also gain the trust of the ES&H professionals and the workers. Therefore, it is important for the leadership team to have developed relationships with other department managers. Gaining trust across departmental lines can pave the way for trust to be developed throughout the company among various levels of the workforce.

14.7 The Corporate Safety Culture and Trust

Culture is an important element in an organization because it influences the behavior of people and guides all aspects of organizational life, to include how people are treated. The definition of culture includes, for the most part, a group of patterns, behaviors, and values that are an inherent part of organization systems or structures. Culture provides the platform in which the organization and its members perform their assigned tasks. The ability of management to successfully implement new business strategies is directly related to how they relate to the culture of that organization. A worker that has trust in the organizational culture is more likely to have the desire to support the vision of the organization.

Theorists have postulated that culture and trust are interrelated. A study was conducted in 2007 that demonstrated there is a strong relationship between organizational culture and organizational trust (Alston, 2007). The study concluded that there is a 0.79% correlation linking the two concepts. Figure 14.3 presents the culture trust integration model (CTIM) that illustrates that both culture and trust share similar attributes that are influential in how workers embrace the organizational environment.

Some of the shared attributes of culture and trust are listed below:

- Open and honest communication
- Treating people with respect and demonstrating integrity

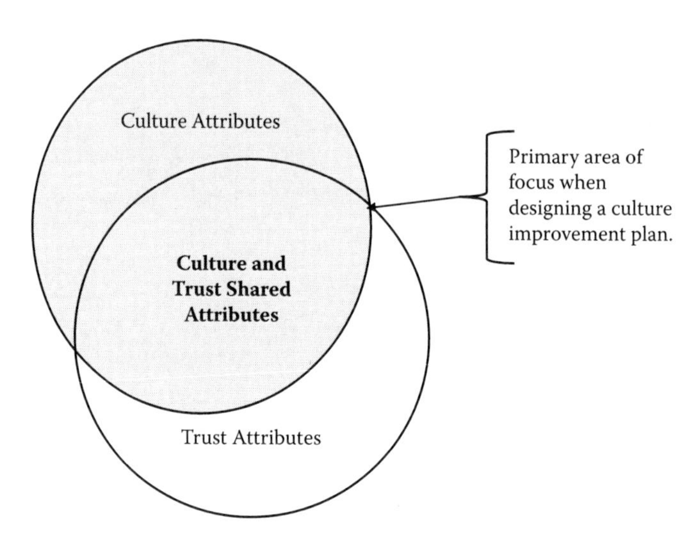

FIGURE 14.3
Culture–trust integration model.

- Management engagement
- Employee involvement
- Policy and procedure implementation
- Accountability

Based on their shared attributes, and that changes in culture can impact trust among people and the organization, it can be postulated that culture directly drives trust. A trusting culture has visual characteristics of accountability and of a learning culture, a culture where people are not blamed for the mistakes they make and use mistakes as an opportunity to learn. In addition, members of the organization are held accountable for actions associated with the decisions that are made.

A very important aspect of culture is the connection that exists between leadership and a trusting culture. The leadership team is the single most important aspect in shaping the culture of an organization. Trust can determine whether a leader is able to gain access to the knowledge and creative thinking possessed by his or her professional staff that are required to solve problems. Management effectiveness may depend on the ability to gain the trust of subordinates and clients. Poor leaders will acknowledge that people are valuable resources, but their *actions* demonstrate something otherwise.

Trust is one of the most important attributes of a thriving corporate culture, and it is important in the effort to build a safety culture that encourages employee engagement. Key attributes of a safety and organizational culture include the following:

- Employees are involved in the decision-making process.
- There is open and honest communication up, down, and across the organization.
- Employees feel free to stop a task if the task is believed to be unsafe to perform.

The culture of an organization is a valuable indicator of the leadership team's ability to tap into the critical thinking skills of the workers, gain support for important initiatives, increase productivity, improve relationships that lead to trust, and retain workers with the skills needed to support the company's mission. Culture is discussed in more detail in Chapter 13.

14.8 Assessing Organizational Trust

Assessments are used across the board by management to gauge performance of key elements at various levels of a process, operation, or organization. Trust is another element that is key to organizational performance, and thereby should be assessed at some frequency. Assessing trust is not performed as one would conduct the traditional process assessment. Therefore, before moving forward with devising a strategy to assess trust within a team, group, or organization, it is critical that the assessor or the assessment team have a good understanding of the appropriate tool or protocol to use to obtain accurate information.

There are many survey tools available that can be used to assess trust in organizations. It is important to select the appropriate survey tool that can provide valuable information on the level of trust within the organization and the element or attributes that are critical for improving trust. Trust can also be assessed through the use of focus group interviews or individual interview sessions. In both cases the questions should be designed to facilitate open and honest feedback from the interviews.

One may ask the question, why assess trust? Assessing trust within your organization provides an indication of the health of the organization, to include relationships between management and the worker, as well as worker-to-worker relationships. It also can provide an indication of the culture and the management team's ability to assess knowledge and creativity of the worker. Below are example questions that may be used when formulating or developing a trust survey:

1. Does senior management in my company communicate information completely and frequently?
2. Do managers in my company support a work-life balance for all workers?

3. Are managers in my organization not afraid to admit when they are wrong?
4. Does clear and concise communication permeate throughout the organization?
5. Do managers in my organization keep the promises they make?
6. Are disagreements and issues addressed in a timely manner?
7. Are leaders in my organization decisive in their decision making?
8. Does management provide clear, concise, and honest feedback to subordinates and colleagues?
9. Do managers in my organization demonstrate good leadership and management skills while conducting business and making decisions?
10. Is information communicated in totality, without holding back critical elements that are important for employees to know?
11. Is my workplace safe and clean?
12. Do people in my organization treat each other with respect?
13. Are organization members willing to share information?
14. Is my company benefit program adequate and comparable to those of similar companies?
15. Do managers in my organization lead with confidence?
16. Do managers in my organization attentively listen to workers?
17. Are organizational members receptive to new ideas?
18. Does communication flow in all directions in my organization to ensure that workers are kept informed?
19. Do people in my organization follow policies and procedures?
20. Do policies and practices in my organization ensure that workers are fairly compensated?
21. Is my organization able to retain skilled workers who complete work?

14.9 Summary

Admittedly trust is becoming harder and harder to achieve and sustain in many organizations. A trustworthy leadership team can begin paving the way to increase trust in their teams, among the employees, and in the organization. Some managers are not investing the effort needed to build trusting relationships and work environments. On the other hand, some managers have been working persistently on building trust with no real sustainable progress.

Workers are finding it difficult to trust their management team in the face of corporate mergers, job losses, and being asked to do more and more with fewer resources available to them. Once trust is lost, it is almost impossible to regain in some cases; therefore, close attention must be paid to culture development and the skills and performance of the leadership team.

References

Alston, F., *Culture and Trust in Technology-Driven Organizations*, CRC Press, Boca Raton, FL, 2013.

Alston, F., The Relationship between Perceived Culture and Trust in Technology-Driven Organizations, dissertation, University of Alabama, Huntsville, 2007.

DOE, *DOE Training Program Handbook: A Systematic Approach to Training*, DOE-HDBK-1078-94, U.S. Department of Energy, Washington, DC, August 1994.

DOE, *Personnel Selection, Training, Qualification, and Certification Requirements for DOE Nuclear Facilities*, DOE Order 426.2, U.S. Department of Energy, Washington, DC, July 2013.

Energy Facility Contractor Group, Assessing Safety Culture in DOE Facilities, EFCOG Meeting Handout, January 23, 2009.

Health Physics Society, Careers in Health Physics, http://www.hps.org/.

Millikin, E.J., and Von Weber, M., Management Assessment—An Integrated Environment, Safety, and Health Management System (ISMS) Core Function for Feedback and Continuous Improvement, presented at Quality Management for Organisational and Regional Development Conference, Palermo, Italy, 2005.

NRC, *Qualification and Training of Personnel for Nuclear Power Plants*, Regulatory Guide 1.8, U.S. Nuclear Regulatory Commission, Washington, DC, May 2000.

Index

10 CFR, 40
  50, 111
  835, 41–42
  850, 56
20 CFR 1910, 18–19
29 CFR, 52, 61
  1910, 54, 62, 67–68
  1910.147, 63
  1926, 62, 67–68
  occupational health requirements of, 71
40 CFR, 5

A

Accountability, 14
ACGIH
  electronic available of regulations, 181
  worker health and safety regulatory requirements, 52
Adverse trends, 142
AEC, 37–38
Aerial lifts, safety guidelines for, 60–62
ALARA design, 6
American Conference of Governmental Industrial Hygienists (ACGIH). *See* ACGIH
American National Standards Institute (ANSI). *See* ANSI
Americans with Disabilities Act (ADA), fitness-for-duty programs, 78
Analytical laboratories, 68
ANSI
  electronic available of regulations, 181
  industrial hygiene standards, 31
  PPE standard, 59
As low as is reasonably achievable (ALARA) design. *See* ALARA design
Asbestos
  industrial hygiene programs for, 53–55
  regulatory guidelines and standards, 54
Asbestos-containing materials and products, 55
Asbestosis, 53–54
Assessment
  criteria and lines of inquiry, 132–133
  ES&H organizations, 125–126
  general information, 132
  identification of assessors, 131–132
  organizational culture, 197–201
  performance-based model of, 127
  plan objectives, 130–131
  reporting and results, 132
  schedule of performance, 131
  techniques and observations, 133–134
  techniques for, 126–127
Assessment programs, radiological protection, 43–44
Audits and inspections, 94–97

B

Beryllium, 55–56
  occupational exposures to, 55
Best practices, 134
Biological agents, definition of, 56
Biosafety, 56
Blasting agent safety, 64
BP *Deepwater Horizon* accident, 47–48
Brazing, 67
Bucket trucks, 60–62
Budget, 152
Building trust, 210
  role of ES&H managers in, 216–217
Business and finance office, 103–104
  functions of, 104

C

Candidates
  communication with, 168–169
  identification of, 167–168
Case management, 102–103
  injury and illness, 80–82

Causation, 76
CDC, biosafety regulations, 56
Centers for Disease Control and Prevention (CDC). *See* CDC
Centralized matrix structure, 25–26
CERCLA, 108
Chemical hazards, medical surveillance of, 75
Chemical safety, 108–110
  requirements, 109
Chemical safety and hygiene programs, 56
Cherry pickers, 60–62
Chronic beryllium disease, 55
Code of Federal Regulations (CFR). *See also* specific titles of
  Title 10, 40
  Title 20, 18–19
  Title 29, 52
  Title 40, 5
Communication, 19
  regulatory compliance and, 212–213
  use of technology for, 182
Communication office, 101–102
Communications plan, culture improvement initiatives, 204
Company management, use of ES&H metrics and performance indicators, 143
Compensability determination, workplace injury/illness, 81–82
Compensatory corrective actions, 142–143
Competence, 215–216
Complacency in the workplace, 46
Compliance, 18–19
  benefits of, 16–17
  penalty of regulatory noncompliance, 17–18
  training requirements of, 112
Compliance auditing, 95–96
Confined spaces, 57
Conflict resolution, 155
Construction safety, 62
  fall protection, 64
Continuous improvement process, use of performance metrics and indicators in, 140–143
Core values, organizational culture and, 193
Corporate safety culture, 217–219
Corrective actions, 142–143
Corrective maintenance, 6
Cost management, 152
Cost–benefit analysis, 190–191
Critical skills worksheet, 167
Crystalline silica, 60
Culture, 193
  attributes of, 203
  environmental protection, 196
  improvement plan, 207
  influencers, 194
  methods to evaluate, 196–201
  radiological protection, 195
  safety and health, 194–195
  tools for improving, 205–207
  trust and, 217–219
Culture improvement initiatives
  communications plan, 204
  defined culture values, 201
  employee involvement committees, 205
  focused campaigns, 204–205
  key performance indicators, 201–202
  leadership and supervisor training, 202–203
Culture–trust integration model, 218
Customers, establishing trust with, 213–216
Cutting, 67

D

Data, electronic validation of, 187–188
Davis–Bessie nuclear incident, 195
Deactivation and demolition activities, 16
Death, company culture and, 195
Deepwater Horizon accident, 47–48
Department of Defense (DoD)
  electronic available of regulations, 181
  reliability and surety programs, 78
Department of Energy (DOE). *See* DOE
Department of Energy Organization Action of 1977, 38

Design process, application of ES&H in, 5–6
Differing opinion resolution, 155
Digital imaging, 186
Dissemination of information, 102
Divisional organization structure, 22–23
Documentation, 6, 105–108
  radiation protection, 45–46
  reviews of for organizational assessment, 126
DOE, 38
  electronic available of regulations, 181
  radiation protection program requirements, 41–43
  reliability and surety programs, 78
  training requirements of, 111–112

E

Electronic data
  security protection of, 188
  validation of, 187–188
Electronic devices
  planning costs associated with, 189
  use of in the workplace, 188
Elevators, 63
Emergency Planning and Community Right-to-Know Act (EPCRA). *See* EPCRA
Emergency response actions, 46–47
  use of technology, 182–183
Emerging trends, 142
Employee involvement committees, culture improvement initiatives, 205
Employee observational programs, 183–184
Employee retention strategy, 162–163
Employees
  benefits of regulatory compliance for, 16–17
  concern for, 215
  feedback from, 8, 120
  involvement in environmental functional area, 93–94
  penalty of regulatory noncompliance for, 18
  perceptions of company culture, 197
Employers, responsibilities of, 13–14
Energy Research and Development Administration (ERDA). *See* ERDA
Environment, benefits of regulatory compliance for, 17
Environment safety and health (ES&H). *See* ES&H
Environmental audits, compliance auditing, 95–96
Environmental functional area (EFA)
  audits and inspections, 94–97
  employee involvement, 93–94
  functions of, 86
  interest groups, 94
  organization structure and design, 85–88
  permitting process, 91–92
  regulatory compliance, 91–92
  regulatory structure and drivers, 88–89
  sustainability, 93
  waste management, 89, 91
Environmental interest groups, 94
Environmental management systems, 95
Environmental permits, 91–92
Environmental protection
  company culture and, 196
  organizational structure of department of, 32–33
  regulatory drivers, 89
Environmental Protection Agency (EPA). *See* EPA
Environmental regulations, 90
Environmental Response, Compensation, and Liability Act (Superfund). *See* Superfund
Environmental science and protection, 11–12
Environmental scientists, roles and responsibilities, 107
Environmental sustainability, 93
EPA, 85
  compliance with, 88–89
  electronic available of regulations, 181
  regulatory drivers for ES&H function, 5 (*See also* 40 CFR)
  training requirements of, 112

EPCRA, 108
  company culture and, 196
ERDA, 38
Ergonomics, 63–64
ES&H
  application of in operations and maintenance activities, 6
  application of in the design process, 5–6
  culture improvement initiatives within, 201–205
  culture in the work environment, 193
  disciplines of, 1
  electronic applications used by, 181–183
  future state of, 13
  hazards identification and control (HI&C) processes, 1–2
  history of, 12
  present state of, 12
  project management approach to, 14–16, 145–146
  regulatory drivers for, 5
  responsibility for, 13–14
  scheduling, 146–152
  training, 34, 120–121
ES&H managers
  building trust, 210
  differing opinion resolution, 155
  establishing trust with customers, 213–216
  leadership skills, 154
  organizational skills, 156
  planning skills, 156–157
  as project managers, 152–154
  resource allocation and management, 156
  role of in building trust, 216–217
  team building, 154–155
  technical knowledge, 156
  time management, 157–158
ES&H organization
  assessing, 125–134
  department or group level, 29
  environmental protection, 32–33
  functional structure, 28
  impact of culture in, 194
  impact of mistrust in, 210–211
  matrix structure, 28–29
  occupational health services, 33
  program support group, 99–100
  radiation protection, 29–30
  role of trust in, 211–213
  schedule example, 149–150
  structure development, 21–26
  structure of, 8–9, 26–27
  succession strategy, 175–177
  support services, 30–31
  use of metrics and performance indicators, 143
  worker safety and health, 31–32
Escalators, 63
Evaluation reports, use of to evaluate training effectiveness, 120
Explosives, 64
Exposures
  asbestos, 53–55
  beryllium, 55
  biological agents, 56
  lead, 57–58
  medical surveillance, 75
  prevention of with PPE, 59
  prevention of with respiratory protection, 60
  validation of data used for reporting, 187–188
External candidates, 175
External hiring process, 175

F

Facility inspection, use of to evaluate training effectiveness, 120
Fall arrest, 64
Fall protection, 64, 66–67
Fall restrain, 64
Field observations
  use of for organizational assessment, 126
  use of to evaluate training effectiveness, 119–120
Findings, 134
Fish and Wildlife Services (FWS), 89
Fitness for duty, 77–78
Focused campaigns, culture improvement initiatives, 204–205
Food and Drug Administration (FDA), food safety, 57

Food safety, 57
Fork trucks, 64–65
Fukushima Daichi nuclear accident, 48
Functional elements of assessments, 127–128
  compliance, 128
  effectiveness, 128–129
  quality, 129
Functional managers. *See* ES&H managers
Functional organization structure, 23–25

G

Gap analysis, 169–170
General duty clause, 5, 18–19
  food safety, 57
  refractory ceramic fiber exposure, 59
Generational differences, 189–190
Graphics, 182
Group interviews, use of to assess company culture, 198–199

H

Hazard communication, 56
Hazard controls, hierarchy of, 4
Hazardous energy, 62–63
  lockout, tagout, 63
Hazardous waste, storage area checklist, 96
Hazards
  identification of, 2–4
  identification of during project lifecycle, 14–16
  mitigation of, 4–5
Hazards identification and control (HI&C) processes, 1–2, 3
  inclusion of in scope of work, 7
  scope of work, 2
Health hazards, chemicals, 56
Health Insurance Portability and Accountability Act (HIPAA). *See* HIPAA
Health service organization
  organization chart, 72
  roles and responsibilities, 107
Health services clinic manager, 33
Hearing conservation, 58
  example program schedule, 151
Heat stress, 57
HIPAA, 81
Hiring process, external, 175
Hoisting, 62
Honesty, 215

I

Identification attribute of trust, 216
Illumination, 65
Individual interviews, use of to assess company culture, 198–199
Industrial hygiene, 10
  regulatory guidelines for, 54
  standards governing, 31
Industrial hygiene programs
  asbestos, 53–55
  beryllium, 55–56
  biosafety, 56
  chemical safety and hygiene, 56
  confined spaces, 57
  definition of, 53
  food safety, 57
  hearing conservation, 58
  heat stress, 57
  lasers, 58–59
  lead, 57–58
  nanotechnology, 58
  personal protective equipment, 59
  refractory ceramic fibers (RCF), 59
  respiratory protection, 60
  silica, 60
Industrial safety, 10–11, 60. *See also* Safety programs
  example programs for, 61
  guidelines, 61
Industrial safety program, 31–32
Injury and illness
  case management and reporting, 80–82, 102–103
  evaluation, 76–77
Internal candidates, 162, 167–168
International Atomic Energy Agency, approaches to company culture, 195

International Organization for Standardization (ISO), 14001 standard, 95
Interviews
  use of for organizational assessment, 126
  use of to assess company culture, 198–199
Ionizing radiation, 9, 37

J

Job analysis, 114, 116
Job performance requirements, training and, 112

K

Key competencies, identification of, 166–167
Key performance indicators (KPIs), 137
  culture improvement initiatives, 201–202
Key positions, 161
  identifying, 165–166

L

Laboratory services, 88
Ladders, 65
Lagging metrics and indicators, 138–139
Laptops, 185
Lasers, 58–59
  control mechanisms, 59
Lead, 57–58
Leadership skills, 154
Leadership team
  role of in succession planning, 163
  succession strategy for, 175–177
Leadership training, culture improvement initiatives, 202–203
Leading metrics and indicators, 138
Learning objectives, developing, 116–117
Lesson plans, 118
Life Safety Code, 181
Lighting, 65
Lines of inquiry, 132–133
Lockout, tagout control, 63

M

Machine guarding, 65
Maintenance activities, application of ES&H in, 6
Management
  benefits of regulatory compliance for, 16
  feedback from, 120
  penalty of regulatory noncompliance for, 17–18
  role of in succession planning, 163
Managing cost, 152
Manhattan Project, 37
Material safety data sheets (MSDSs), 108–109
  chemical hazard communication, 56
Matrix organization structure, 25–26
  worker safety and health department, 51–52
McMahon/Atomic Energy Act, 37
Media, organization and storage of, 187
Medical personnel
  challenges of managing, 82–83
  functions of in occupational health program, 73–79
  videoconferencing with, 182
Medical staff, alignment of goals with, 72–73
Medical surveillance, 11, 67, 75–76
Mentors, role of in succession planning, 174–175
Mesothelioma, 53–54
Mission statement, organizational culture and, 193
Mistrust, impact of in an organization, 210–211
Monitoring trends, 142

N

Nanotechnology, 58
National Electric Code (NEC), worker health and safety regulatory requirements, 52
National Fire Protection Association (NFPA). *See* NFPA
National Institute of Occupational Safety and Health (NIOSH). *See* NIOSH

National Marine Fisheries Service (NMFS), 89
Needs analysis, 114–115
Networking, 188–189
NFPA
  electronic available of regulations, 181
  industrial hygiene standards, 31
  worker health and safety regulatory requirements, 52
NIOSH
  electronic available of regulations, 181
  worker health and safety regulatory requirements, 52
Noise, 58
Noncompliance, 91–92
  penalty of, 17–18
Nonionizing radiation, 9
Nonregulatory assessment criteria, 128–129
NRC, 38
  approaches to company culture, 195
  electronic available of regulations, 181
  fitness-for-duty programs, 78
  radiation protection program requirements, 41–43
  regulatory drivers for ES&H function, 5
  training requirements of, 111–112
Nuclear accidents
  Deepwater Horizon, 47–48
  Fukushima Daichi, 48
Nuclear facilities, emergency response at, 46–47
Nuclear Regulatory Commission (NRC). *See* NRC

O

Objective-based training, development of, 116–117
Observational programs, 183–184
Observations, 134
Occupational exposure limits (OELs), 67–68
Occupational exposures
  beryllium, 55
  biological agents, 56
  chemicals, 56
  confined spaces, 57
  heat stress, 57
  lasers, 58–59
  lead, 57–58
  medical surveillance, 75
  nanotechnology, 58
  noise, 58
  refractory ceramic fibers (RCF), 59
  silica, 60
  validation of data used for reporting, 187–188
Occupational health, 11
  definition of, 71
Occupational health program
  challenges of managing, 82–83
  facets of, 71
  functions of, 73–79
  illness and injury case management, 80–82
  management and administration, 72–73
  prevention and wellness, 79–80
Occupational health services, organizational structure of department of, 33
Occupational safety, 60. *See also* Safety programs
Occupational Safety and Health Act of 1970, 12
Occupational Safety and Health Administration (OSHA). *See* OSHA
One-on-one daily discussions, 197
Openness, 215
Operations, application of ES&H in, 6
Opportunities, 134
Organization design, 8–9
  environmental science and protection, 11–12
  occupational health, 11
  radiation safety, 9
  worker safety and health (WS&H), 10–11
Organization structure
  designing for company size, 27
  development, 21–22
  divisional, 22–23
  environmental protection group, 32–33

ES&H functional organization, 28
ES&H matrix organization, 28–29
ES&H training department, 34
functional, 23–25
matrix, 25–26
occupational health services department, 33
radiation protection department, 29–30
support services group, 30–31
worker safety and health department, 31–32
Organizational culture, 193
environmental protection, 196
methods to evaluate, 196–201
radiological protection, 195
safety and health, 194–195
tools for improving, 205–207
trust and, 217–219
Organizational culture questionnaire, 206–207
Organizational skills, 156
enhancers for, 157
Organizational trust. *See also* Trust
assessing, 219–220
OSHA, 12
asbestos guidelines, 54
beryllium requirements, 56
compliance with regulations of, 18–19
construction safety standards, 62
electronic available of regulations, 181
general duty clause, 5, 18–19, 57, 59
Hazard Communication Standard, 56
industrial hygiene standards, 31
machine guarding standards, 65
permissible exposure limits (PELs), 67–68
regulations for PPE, 59
regulatory drivers for ES&H function, 5
regulatory reporting determinations, 76
reportability requirements, 80
worker health and safety regulatory requirements, 52

P

Penalties of noncompliance, 17–18, 91–92
Performance assurance office, 100–101
activities of, 101
Performance indicators, 135, 137–138
lagging, 138–139
leading, 138
qualities of, 139–140
use of to assess company culture, 200–201
Performance metrics, 134–136
initiatives, 204
lagging, 138–139
leading, 138
qualities of, 139–140
use of in continuous improvement process, 140–143
Performance metrics and indicator improvement model (PMIIM), 140
Performance of work, 7
Performance tests, 117
Performance-based assessment
model, 127–130
plan objectives, 130–131
use of functional elements in, 128–130
Permissible exposure limits (PELs), 67–68
Permitting process, 91–92
Personal protective equipment (PPE), 59
Physical exams
fitness-for-duty, 77
preemployment, 74–75
return-to-work process, 78–79
routine, 75–76
Planning skills, 156–157
Planning stage, 14
Policies and procedures, 105
Portable instrumentation, 185–186
Powered industrial trucks, 64–65
Preemployment physicals, 74–75
laws associated with, 74
Pressure vessels, 66
Prevention and wellness programs, 79–80
Preventive corrective actions, 143
Preventive maintenance, 6
Program management office, 104–108
activities of, 105

Program support group, 99–100
- activities of, 100
- business and finance office, 103–104
- chemical safety, 108–110
- communication office, 101–102
- injury and illness management and reporting office, 102–103
- performance assurance office, 100–101
- program management office, 104–108

Project closure, 16
Project execution, 15
Project initiation, 15
Project management
- approach to ES&H, 14–16
- differing opinion resolution, 155
- ES&H scheduling, 146–152
- leadership skills, 154
- managing cost, 152
- resource allocation and management, 156
- skills, 153
- team building, 154–155

Project management approach, 145–146
Project managers
- ES&H managers as, 152–154
- organizational skills, 156
- planning skills, 156–157
- technical knowledge, 156
- time management, 157–158

Project schedule, 146–152
- sequencing of, 148

Project stages and tasks, 147
Publications, worker health and safety, 52–54

R

R2A2s, 13–14
Radiation protection
- balance between safety and production, 47–48
- company culture and, 195
- complacency in the workplace, 46
- documentation, 45–46
- management and organizational structure, 38
- organization structure of department of, 29–30
- policies and procedures, 42–43
- program assessment, 43–44
- program functional elements, 43
- programs, 41–43
- regulatory requirements, 40
- response to abnormal conditions, 46–47
- roles and responsibilities, 39, 106
- training in, 44–45

Radiation safety, 9
Radiological control manuals, elements of, 41–42
Radiological control officer (RCO), responsibilities of, 39
Radiological incidents, reportability requirements, 80
Radiological program, flow-down of requirements, 39–40
Radiological protection professionals, 38
- emergency response actions of, 46–47

Refractory ceramic fibers (RCF), 59
Regulations, 11–12
- structure and drivers of environmental function, 88–89
- use of technology, 180–181
- worker health and safety, 52–54

Regulatory compliance, 91–92
- benefits of, 16–17
- communication and, 212–213
- training requirements of, 112

Regulatory requirements, radiation protection, 40
Relationship building, 215
Reliability, 216
Remedial corrective actions, 143
Reporting, injury and illness, 80–82
Reporting structure, 8–9
Resource allocation and management, 156, 163
Respiratory protection, 60
Retention strategy, 162–163
Return-to-work process, 78–79
Rigging, 62

Roles, responsibility, accountability, and authority. *See* R2A2s
Routine medical physicals, 75–76

S

Safety data sheets (SDSs), chemical hazard communication, 56
Safety programs
aerial lifts, 60–62
brazing, 67
construction safety, 62
cutting, 67
design of, 60
elevators and escalators, 63
ergonomics, 63–64
explosives and blasting agents, 64
fall protection, 64
hazardous energy, 62–63
hoisting and rigging, 62
ladders, 65
lighting and illumination, 65
machine guarding, 65
powered industrial trucks, 64–65
pressure vessels, 66
signage, 66
soldering, 67
traffic, 66
walking-working surfaces, 66–67
welding, 67
SARA, 108
company culture and, 196
Schedules, guidelines for developing, 148
Scheduling, 146–152
benefits of using, 147
sequencing of, 148
Scope of work, 7
definition of for hazards identification and control, 2
Search engines, 180
Signs, 66
Silica, 60
Simulation devices, 186
Site occupational medicine doctor (SOMD), 33
Skill gap analysis, 169–170
Slips, 66–67
Smartphones, 184–185
Social media, 188–189
Soldering, 67
Staff development, 161
Stakeholders
benefits of regulatory compliance for, 17
penalty of regulatory noncompliance for, 18
Standards
use of technology, 180–181
worker health and safety, 52–54
Succession plan
attributes of, 164–165
evaluation of, 173–174
Succession planning
candidate communication, 168–169
defining the training and development plan, 170–173
definition of, 161–162
identification of candidates, 167–168
identification of key competencies, 166–167
identification of key positions, 165–166
role of a mentor in, 174–175
role of management in, 163
training and development plan, 170–174
Succession strategy, ES&H organizational, 175–177
Superfund, 89
Superfund Amendments and Reauthorization Act (SARA). *See* SARA
Supervisors
culture improvement initiatives, 202–203
feedback from, 120
Support services, organization structure of department of, 30–31
Surveys
use of for organizational assessment, 127
use of to assess company culture, 199–200
Sustainability, 93
Systems approach to training (SAT)
analysis, 114–116
design of training, 116–117
development of training, 117–118

evaluation for effectiveness, 119–120
implementation of training, 118–119
regulatory compliance and, 111–113

T

Tablets, 185
Task analysis, 115
Team building, 154–155
Team cohesion model, 155
Technology
applications used by ES&H discipline, 181–182
communications and graphics, 182
considerations, 187–190
cost–benefit analysis, 190–191
emergency response, 182–183
employee observational programs, 183–184
ES&H professionals and, 179
knowledge requirements for, 156
overcoming generational differences, 189–190
planning costs associated with, 189
regulations and standards, 180–181
search engines, 180
training and certification programs, 184
types of devices, 184–186
use of in the workplace, 180
Terminal learning objectives, developing, 116–117
Test questions, developing, 117
Time management, 157–158
Total recordable case (TRC) rate, 135, 139
organizational culture and, 194–195
Traffic safety, 66
Training, 111–113
analysis of, 114–116
design of, 116–117
development of, 117–118
ES&H, 120–121
evaluation for effectiveness, 119–120
hazardous energy, 63
implementation of, 118–119
medical personnel, 73
method and application, 119
organization structure of department, 34
powered industrial trucks, 64–65
radiation protection, 44–45
records retention, 122
systematic approach to, 113
technology, 189
tracking of, 120–122
use of technology for, 184
worker safety and health program, 51–52
Training and development plan
defining, 170–173
implementation of, 173
Training and evaluations standards, 116–117
Transparency, 14
Trending, 141–142
Trips, 66–67
Trust, 209
assessing, 219–220
building, 210
competence, 215–216
concern for employees, 215
corporate safety culture and, 217–219
establishing with customers, 213–214
humility and, 215
identification attribute of, 216
impact of mistrust in an organization, 210–211
openness and honesty, 215
relationship building, 215
reliability, 216
role of ES&H managers in building, 216–217
role of in an ES&H organization, 211–213
Trust paradigm, 214
Trust-sustaining model, 214

U

U.S. Army Corps of Engineers (USACE), 89
U.S. Atomic Energy Commission (AEC). *See* AEC
U.S. Department of Agriculture, food safety, 57
U.S. Department of Defense. *See* Department of Defense (DoD)

U.S. Department of Energy (DOE). *See* DOE
Union Carbide, 196

V

Videoconferencing, 182

W

Walking-working surfaces, 66–67
Waste management, 89, 91
Watch dog groups, 94
Welding, 67
Wellness programs, 79–80
Work control processes, 6–7
Work restrictions, 79
Work-related injuries
  ergonomics and, 63–64
  falls, 64
Worker feedback, 8
Worker safety and health (WS&H), 10
  industrial hygiene, 10
  industrial safety, 10–11
  occupational exposure limits, 68
  organization structure of department of, 31–32, 51–52
  organizational culture and, 194–195
  program design, 52
  program goals, 51
  replacing functional managers, 176
  roles and responsibilities, 106
Workers' compensation, 103
  intent of, 81–82
Workplace complacency, 46
Workplace injuries, company culture and, 195
Workplace monitoring, 75
Workplace surveys, use of for organizational assessment, 127